SpringerBriefs in Comput

MW00837661

Series Editors
Stan Zdonik
Shashi Shekhar
Jonathan Katz
Xindong Wu
Lakhmi C. Jain
David Padua
Xuemin (Sherman) Shen
Borko Furht
V.S. Subrahmanian
Martial Hebert
Katsushi Ikeuchi
Bruno Siciliano
Sushil Jajodia
Newton Lee

More information about this series at http://www.springer.com/series/10028

Xun Yi • Russell Paulet • Elisa Bertino

Homomorphic Encryption and Applications

 Springer

Xun Yi
RMIT University
Computer Science & Info Tech
Melbourne, VIC, Australia

Russell Paulet
Victoria University
Melbourne, VIC, Australia

Elisa Bertino
Computer Science
Purdue University
West Lafayette, IN, USA

ISSN 2191-5768   ISSN 2191-5776 (electronic)
ISBN 978-3-319-12228-1   ISBN 978-3-319-12229-8 (eBook)
DOI 10.1007/978-3-319-12229-8
Springer Cham Heidelberg New York Dordrecht London

Library of Congress Control Number: 2014953221

Printed on acid-free paper

Springer is part of Springer Science+Business Media (www.springer.com)

To our families

Preface

Homomorphic encryption is a form of encryption that allows specific types of computations to be carried out on ciphertext and generate an encrypted result that, when decrypted, matches the result of operations performed on the plaintext.

This is a desirable feature in modern communication system architectures. The homomorphic property of various cryptosystems can be used to create secure voting systems and private information retrieval schemes and enable widespread use of cloud computing by ensuring the confidentiality of processed data.

This book presents the basic homomorphic encryption techniques and their applications. It begins with an introduction of the history of encryption techniques from classical ciphers to secret key encryption and public-key encryption, including secret key encryption and public-key encryption models. It then provides the definition of homomorphic encryption followed by the description of some well-known homomorphic encryption schemes, such as the ElGamal and Paillier encryption schemes. On the basis of the homomorphic encryption concept, this book further introduces the state-of-the-art fully homomorphic encryption concept and describes the fully homomorphic encryption schemes over integers. After that, this book focuses on three applications of homomorphic encryption techniques. The first application introduces an electronic voting scheme on the basis of the ElGamal encryption scheme. The second application deals with nearest neighbor queries with location privacy on the basis of private information retrieval built on the Paillier encryption scheme. The third application discusses private searching on streaming data on the basis of fully homomorphic encryption schemes.

This book is designed to serve as a reference book for undergraduate- or graduate-level courses in computer science or mathematics departments, as a general introduction suitable for self-study (especially for beginning graduate students), and as a reference for students, researchers, and practitioners.

RMIT University, Melbourne, VIC, Australia   Xun Yi
Victoria University, Melbourne, VIC, Australia   Russell Paulet
Purdue University, West Lafayette, IN, USA   Elisa Bertino
September 2014

Acknowledgments

We would like to express our appreciation to Professor Udaya Parampalli (The University of Melbourne, Australia) and Dr. Junzuo Lai (Jinan University, China) for their comments on our book.

Contents

Chapter 1
Introduction

Abstract Encryption is the process of converting messages, information, or data into a form unreadable by anyone except the intended recipient. Encrypted data must be decrypted, before it can be read by the recipient. In its earliest form, people have been attempting to conceal certain information that they wanted to keep to their own possession by substituting parts of the information with symbols, numbers, and pictures. Today's encryption algorithms are divided into two categories: secret key and public key. Secret key encryption schemes use the same key (the secret key) to encrypt and decrypt a message, and public-key encryption schemes use one key (the public key) to encrypt a message and a different key (the private key) to decrypt it, and all of today's encryption algorithms fit within those two categories. This chapter introduces the history of encryption techniques from classical ciphers to secret key encryption and public-key encryption, including secret key and public-key encryption models. It provides some background for homomorphic encryption.

1.1 Classical Ciphers

A cipher is a technique for hiding a message, by which letters of the message are substituted or transposed to other letters, letter pairs, and even many letters. In cryptography, a classical cipher is a type of cipher that was used historically but not now. In general, classical ciphers operate on an alphabet of letters (such as "A–Z") and can be implemented by hand or with simple mechanical devices. They are the most basic types of ciphers and not very secure, especially after new technology was developed. Modern schemes use computers or other digital technology and operate on bits and bytes.

Many classical ciphers were used by well-respected people, such as Julius Caesar and Napoleon, who created their own ciphers which were then popularly used. Many ciphers had their origins in the military and were used for transporting secret messages among people on the same side.

Classical schemes are often susceptible to ciphertext-only attacks, sometimes even without knowledge of the encryption system itself, using tools such as frequency analysis.

Classical ciphers are often divided into substitution ciphers, transposition ciphers, and product ciphers as follows:

© Xun Yi, Russell Paulet, Elisa Bertino 2014

X. Yi et al., *Homomorphic Encryption and Applications*, SpringerBriefs in Computer Science, DOI 10.1007/978-3-319-12229-8_1

1. Substitution cipher is a method of encryption by which plaintext letters are replaced with ciphertext letters, according to an encryption system. The receiver decrypts the ciphertexts by performing an inverse substitution.
2. Transposition cipher is a method of encryption by which the positions held by plaintext letters are shifted according to an encryption system, so that the ciphertext letters constitute a permutation of the plaintext letters. Mathematically a bijective map is used on the letters' positions to encrypt and an inverse map to decrypt.
3. Product cipher combines a sequence of simple transformations such as transposition ciphers and substitution ciphers. The combination could yield a cipher system more powerful than either one alone.

1.1.1   Substitution Ciphers

A well-known example of substitution ciphers is the Caesar cipher [14]. The Caesar cipher is named after Julius Caesar (July 100 BC–15 March 44 BC), who was a Roman general, statesman, and Consul and played a critical role in the events that led to the demise of the Roman Republic and the rise of the Roman Empire. Caesar was the first recorded use of this cipher.

To encrypt a message with the Caesar cipher, each letter of message is replaced by the letter three positions later in the alphabet. Hence, A is replaced by D, B by E, C by F, etc. Finally, X, Y, and Z are replaced by A, B, and C respectively. So, for example, "CAESAR" encrypts as "FDHVDU." Caesar rotated the alphabet by three letters, but any number works. When the number of rotations is 19, the plaintext and ciphertext alphabets look like:

Plaintext alphabet: $a\ b\ c\ d\ e\ f\ g\ h\ i\ j\ k\ l\ m\ n\ o\ p\ q\ r\ s\ t\ u\ v\ w\ x\ y\ z$
Ciphertext alphabet: $t\ u\ v\ w\ x\ y\ z\ a\ b\ c\ d\ e\ f\ g\ h\ i\ j\ k\ l\ m\ n\ o\ p\ q\ r\ s$

While the encryption is done by substituting plaintext letters with the corresponding ciphertext letters, the decryption is done by performing an inverse substitution. The encryption and decryption processes can be implemented by a cipher wheel as shown in Fig. 1.1, where the plaintext and ciphertext alphabets are on the outer and inner wheels, respectively, and the inner wheel is turnable.

As there are only 25 possible rotations for the alphabet, the Caesar cipher can be easily broken by a brute-force attack or exhaustive key search, i.e., systematically checking all possible keys until the correct one is found.

The Caesar cipher is a monoalphabetic substitution cipher, where just one ciphertext alphabet is used. It is also possible to have a polyalphabetic substitution cipher, where multiple ciphertext alphabets are used. This makes the ciphertext much harder to decode because the codebreaker would have to figure out ciphertext alphabets used.

Fig. 1.1 Caesar cipher wheel

A typical example of polyalphabetic substitution ciphers is the Vigenere cipher [5]. The Vigenere cipher is named after Blaise de Vigenere (5 April 1523–19 February 1596), who was a French diplomat, cryptographer, translator, and alchemist.

To encrypt, a table of alphabets, as shown in Fig. 1.2, is used, termed Vigenere square, composed of the alphabet written out 26 times in different rows, each alphabet shifted cyclically to the left compared to the previous alphabet, corresponding to the 26 possible Caesar ciphers.

To use the Vigenere square to encrypt a message, we first choose a keyword and then repeat it until it is the same length as the message we wish to encode. We then would write the message underneath the repeated keyword to see which ciphertext alphabet you would use for each letter of the message. The first letter of the message would be encrypted using the ciphertext alphabet that corresponds with the first letters of the keyword. For example if we have a keyword of VENUS and the message we want to encode is polyalphabetic, this is what we would do:

Keyword: *V E N U S V E N U S V E N U*
Plaintext: *p o l y a l p h a b e t i c*
Ciphertext: *K S Y S S G T U U T Z X V W*

Some substitution ciphers involve using numbers instead of letters. An example of this is the Great Cipher [21], where numbers were used to represent syllables.

1.1.2 Transposition Ciphers

In a transposition cipher, the letters themselves are kept unchanged, but their order within the message is scrambled according to some well-defined scheme. Many transposition ciphers are done according to a geometric design. A simplest example of transposition ciphers is the Scytale cipher [13]. The Scytale cipher was used by the ancient Greeks and Spartans to communicate during their military campaigns. It was first mentioned by the Greek poet Archilochus, who lived in the seventh century BC. The Scytale cipher involves three pieces of equipment, namely a pen, a

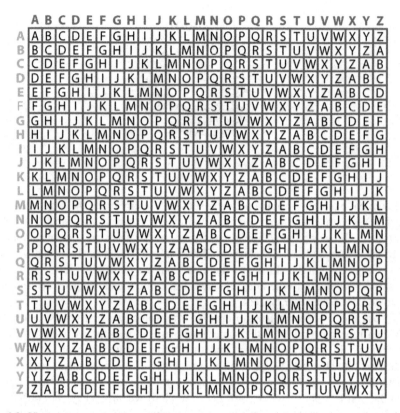

Fig. 1.2 Vigenre square

long strip of paper (leather was used by the Greeks and Spartans), and a cylinder of some sort, as shown in Fig. 1.3.

The long thin strip of paper is then wrapped around the cylinder, going from one end to the other. The message

KILL KING TOMORROW MIDNIGHT

is then written horizontally on the paper, one letter for each wrap around, going from left to right, three letters per column. The cylinder is rotated and the rest of the message is written until the message is complete. Once its complete, the strip of paper is taken off and the result

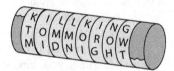

Fig. 1.3 Scytale

$$K\,T\,M\,I\,O\,I\,L\,M\,D\,L\,O\,N\,K\,R\,I\,I\,R\,G\,N\,O\,H\,G\,W\,T$$

is the ciphertext. With this ciphertext, the only way to read the original is to re-wrap it around a cylinder of equal width and read the letters from left to right.

The diameter of the Scytale can be regarded as the key of the cipher. Since the key can take a limited positive integer only, the Scytale cipher can be easily broken by a brute-force attack.

Another simple example of transposition ciphers is the columnar cipher [12]. It can be performed by hand. First, the message is written out in rows of a fixed length, and then read out again column by column, and the columns are chosen in some scrambled order. Both the width of the rows and the permutation of the columns are usually defined by a keyword. For example, suppose we use the keyword GERMAN and the message

$$defend\ the\ east\ wall\ of\ the\ castle.$$

The encryption process can be illustrated as follows:

G	E	R	M	A	N
d	e	f	e	n	d
t	h	e	e	a	s
t	w	a	l	l	o
f	t	h	e	c	a
s	t	l	e	x	x

$\longrightarrow$

A	E	G	M	N	R
n	e	d	e	d	f
a	h	t	e	s	e
l	w	t	l	o	a
c	t	f	e	a	h
x	t	s	e	x	l

In the above example, the plaintext has been padded with "xx" so that it neatly fits in a rectangle. This is known as a regular columnar transposition. An irregular columnar transposition leaves these characters blank, though this makes decryption slightly more difficult. The columns are now reordered such that the letters in the keyword are ordered alphabetically. The ciphertext is read off along the columns, i.e.,

$$nalcxehwttdttfseeleedsoaxfeahl$$

Many transposition ciphers are similar to these two examples, usually involving rearranging the letters into rows or columns and then taking them in a systematic way to transpose the letters.

1.1.3   Product Ciphers

A product cipher combines two or more transformations in a manner intending that the resulting cipher is more secure than the individual components. A typical

	A	**D**	**F**	**G**	**V**	**X**
A	D	H	X	M	U	4
D	P	3	J	6	A	O
F	I	B	Z	V	9	W
G	1	N	7	0	Q	K
V	F	S	L	Y	C	8
X	T	R	5	E	2	G

example of product ciphers is the ADFGVX cipher [12]. The ADFGVX cipher was used by the German army during World War I. Invented by Colonel Fritz Nebel and introduced in March 1918, the cipher was combined with a substitution cipher and a transposition cipher. The cipher is named after the six possible letters used in the ciphertext: A, D, F, G, V, and X. These letters were chosen deliberately because they sound very different from each other when transmitted via morse code. The intention was to reduce the possibility of operator error.

The ADFGVX cipher used a 6×6 matrix to substitution-encrypt the 26 letters and 10 digits into pairs of the symbols A, D, F, G, V, and X. The resulting biliteral cipher was then written into a rectangular array and route encrypted by reading the columns in the order indicated by a keyword.

The "key" for a ADFGVX cipher is a "key square" and a key word. The key square is a 6 by 6 square containing all the letters and the numbers 0–9 as shown in Fig. 1.4. The keyword is any word, e.g., GERMAN.

There are a number of steps involved:

1. Build a table like that shown in Fig. 1.4 as the key square. This is known as a Polybius square.
2. Encode the plaintext using this matrix; to encode the letter "a," locate it in the matrix and read off the letter on the far left side on the same row, followed by the letter at the top in the same column. In this way each plaintext letter is replaced by two cipher text letters, e.g., "attack" is encrypted to

DV XA XA DV VV GX

The ciphertext is now twice as long as the original plaintext. Note that so far, it is just a simple substitution cipher and trivial to break.

3. Write the keyword with the enciphered plaintext underneath, e.g.,

G	E	R	M	A	N
D	V	X	A	X	A
D	V	V	V	G	X

4. Perform a columnar transposition. Sort the keyword alphabetically, moving the columns as you go. Note that the letter pairs that make up each letter get split apart during this step; this is called fractionating.

A	E	G	M	N	R
X	V	D	A	A	X
G	V	D	V	X	V

Read the final ciphertext off in columns.

XGVVDDAVAXXV

In the days of manual cryptography, product ciphers were a useful device for cryptographers, and in fact double transposition or product ciphers on keyword-based rectangular matrices were widely used. There was also some use of a class of product ciphers known as fractionation systems, wherein a substitution was first made from symbols in the plaintext to multiple symbols (usually pairs, in which case the cipher is called a biliteral cipher) in the ciphertext, which was then encrypted by a final transposition, known as superencryption.

The great French cryptanalyst Georges J. Painvin succeeded in cryptanalyzing critical ADFGVX ciphers in 1918 [16], with devastating effect for the German army at the Second Battle of the Marne.

Nowadays, most of classical ciphers have become less popular. They were frequently used during World War II, but since computer have become available to security analysis, their applicability has diminished. However, this does not imply that a description of classical ciphers is only of historical interest. These ciphers have had a profound impact on today's information security technology and provide an approach for beginners to understand ideas of information security technology.

1.2 Secret Key Encryption

1.2.1 Secret Key Encryption Model

Secret key encryption algorithms are a class of algorithms that use the same secret keys for both encryption of plaintext and decryption of ciphertext. The keys may be identical or there may be a simple transformation to go between the two keys. The keys, in practice, represent a shared secret between two or more parties that can be used to maintain a private information link.

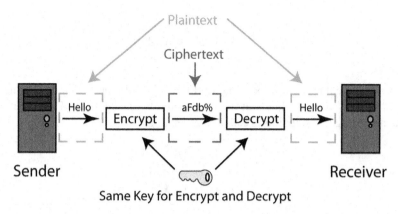

Fig. 1.5 Secret key encryption model

A first systematic and information-theoretic study of secret key cryptosystem can be found in Shannon's classical paper "Communication Theory of Secrecy Systems" [20]. This paper was the first to introduce a secret key encryption model, as shown in Fig. 1.5.

Prior to transmission of a plaintext P, a key source provides both the sender and the recipient with a shared key K. This key is used by the sender to encrypt the plaintext, obtaining a ciphertext C which is delivered to the receiver and possibly intercepted by an enemy eavesdropper. The receiver then uses the key K in order to reconstruct the clear plaintext P.

1.2.2 Data Encryption Standard

The data encryption standard (DES) is a secret key cryptosystem for the encryption of electronic data [19]. It was developed in the early 1970s at IBM and is based on an earlier design by Horst Feistel. The algorithm was submitted to the National Bureau of Standards (NBS) following the agency's invitation to propose a candidate for the protection of sensitive, unclassified electronic government data. In 1976, after consultation with the National Security Agency (NSA), the NBS eventually selected a slightly modified version, which was published as an official federal information processing standard (FIPS) for the United States in 1977.

The overall structure of DES is shown in Fig. 1.6. There are 16 identical stages of processing, termed rounds. There is also an initial and final permutation, termed IP and FP, which are inverses. Before the main rounds, the block is divided into two 32-bit halves and processed alternately; this criss-crossing is known as the Feistel scheme. The Feistel structure ensures that decryption and encryption are very similar processes—the only difference being that the subkeys are applied in the reverse order when decrypting. The rest of the algorithm is identical. This greatly

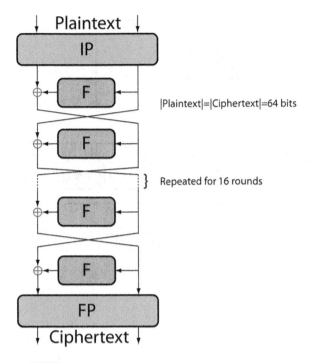

Fig. 1.6 Structure of DES

simplifies implementation, particularly in hardware, as there is no need for separate encryption and decryption algorithms.

The ⊕ symbol denotes the exclusive-OR (XOR) operation. The F-function scrambles half a block together with some of the key. The output from the F-function is then combined with the other half of the block, and the halves are swapped before the next round. After the final round, the halves are swapped; this is a feature of the Feistel structure which makes encryption and decryption similar processes.

The F-function, depicted in Fig. 1.7, operates on half a block (32 bits) at a time and consists of four stages:

- Expansion—the 32-bit half-block is expanded to 48 bits using the expansion permutation, denoted E in the diagram, by duplicating half of the bits. The output consists of eight 6-bit (8 × 6 = 48 bits) pieces, each containing a copy of 4 corresponding input bits, plus a copy of the immediately adjacent bit from each of the input pieces to either side.
- Key mixing—the result is combined with a subkey using an XOR operation. 16 48-bit subkeys—one for each round—are derived from the main key using the key schedule (described below).

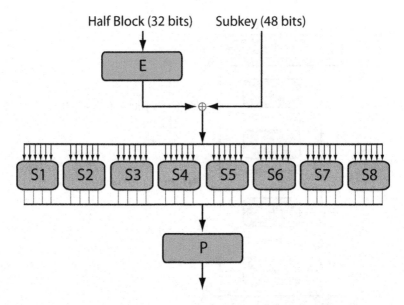

Fig. 1.7 F-function

- Substitution—after mixing in the subkey, the block is divided into eight 6-bit pieces before processing by the S-boxes, or substitution boxes. Each of the eight S-boxes replaces its six input bits with four output bits according to a nonlinear transformation, provided in the form of a lookup table. The S-boxes provide the core of the security of DES—without them, the cipher would be linear and trivially breakable.
- Permutation—finally, the 32 outputs from the S-boxes are rearranged according to a fixed permutation, the P-box. This is designed so that, after permutation, each S-box's output bits are spread across 4 different S-boxes in the next round.

The alternation of substitution from the S-boxes and permutation of bits from the P-box and E-expansion provide the so-called confusion and diffusion, respectively, a concept identified by Claude Shannon in the 1940s as a necessary condition for a secure yet practical cipher. Diffusion means that if we change a character of the plaintext, then several characters of the ciphertext should change, and similarly, if we change a character of the ciphertext, then several characters of the plaintext should change. Confusion means that the key does not relate in a simple way to the ciphertext. In particular, each character of the ciphertext should depend on several parts of the key.

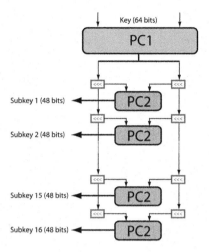

Fig. 1.8 Key schedule of DES

The key schedule of DES is illustrated in Fig. 1.8. It generates 16 subkeys.

Initially, 56 bits of the key are selected from the initial 64 bits by Permuted Choice 1 (PC-1) and the remaining eight bits are either discarded or used as parity check bits. The 56 bits are then divided into two 28-bit halves; each half is thereafter treated separately. In successive rounds, both halves are rotated left by one or two bits (specified for each round), and then 48 subkey bits are selected by Permuted Choice 2 (PC-2), 24 bits from the left half and 24 from the right. The rotations (denoted by "$<<<$" in the diagram) mean that a different set of bits is used in each subkey; each bit is used in approximately 14 out of the 16 subkeys.

The key schedule for decryption is similar. The subkeys are in reverse order compared to encryption. Apart from that change, the process is the same as for encryption.

Although more information has been published on the cryptanalysis of DES than any other block cipher, the most practical attack to date is still a brute-force approach. There are three attacks known that can break the full 16 rounds of DES with less complexity than a brute-force search: differential cryptanalysis (DC) [2], linear cryptanalysis (LC) [15], and Davies' attack [9]. However, the attacks are theoretical and are unfeasible to mount in practice.

1.2.3 Advanced Encryption Standard

The DES cryptosystem (with its variations) was widely used for more than 20 years. The main problem of the DES algorithm was its relatively short secret key, with 2^{56} possible keys. Although this is a fairly large number, with sufficient computational resources brute-force attacks on DES are feasible. So-called DES challenges, where

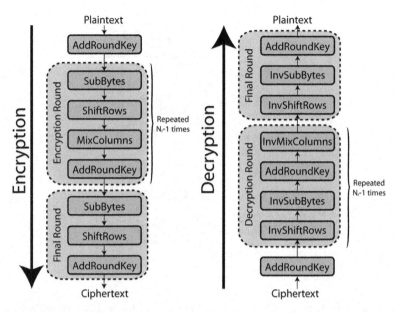

Fig. 1.9 Encryption and decryption of AES

a large number of computers connected to the Internet exhaustively searched the key space, demonstrated this weakness dramatically. The first DES challenge in 1997 was completed in 4.5 months, the second in 1998 in 39 days, and the third and final DES challenge in 1999 was completed in less than a day (22.5 h).

In 1997 the US National Institute of Standards and Technology (NIST) started a public competition to select an algorithm to replace DES. The algorithm was required to support key sizes of 128, 192, and 256 and to be free of any patents. The selection process consisted of several rounds where candidate algorithms were evaluated. At the end of the first round in August 1998, 15 algorithms were accepted as candidates. In the next round in August 1999, the candidates were reduced to five finalist algorithms (MARS, Blowfish, RC6, Rijndael, Serpent). Finally, in April 2000 the Rijndael algorithm was selected as the winner. On 2 October 2000, NIST officially announced that Rijndael has been chosen as Advanced Encryption Standard (AES) [8].

The AES algorithm operates on 128-bit data blocks supporting three different key sizes of 128, 192, and 256 bits. These three flavors of the AES algorithm are also referred to as AES-128, AES-192, and AES-256, for 128-, 192-, and 256-bit keys, respectively. An AES encryption process consists of a number of encryption rounds (N_r) that depends on the size of the key. The standard calls for 10 rounds for AES-128, 12 rounds for an AES-192, and 14 rounds for an AES-256.

During encryption, each round is composed of a set of four basic operations. The decryption process applies the inverse of these operations in reverse order. Figure 1.9 shows the basic structure of the AES encryption and decryption.

$$\begin{array}{|c|c|c|c|}
\hline
S_{0,0} & S_{0,1} & S_{0,2} & S_{0,3} \\
\hline
S_{1,0} & S_{1,1} & S_{1,2} & S_{1,3} \\
\hline
S_{2,0} & S_{2,1} & S_{2,2} & S_{2,3} \\
\hline
S_{3,0} & S_{3,1} & S_{3,2} & S_{3,3} \\
\hline
\end{array}$$

Fig. 1.10 State of AES

$$\begin{bmatrix} b_0 \\ b_1 \\ b_2 \\ b_3 \end{bmatrix} = \begin{bmatrix} 2 & 3 & 1 & 1 \\ 1 & 2 & 3 & 1 \\ 1 & 1 & 2 & 3 \\ 3 & 1 & 1 & 2 \end{bmatrix} \begin{bmatrix} a_0 \\ a_1 \\ a_2 \\ a_3 \end{bmatrix}$$

Fig. 1.11 MixColumns of AES encryption

$$\begin{bmatrix} r_0 \\ r_1 \\ r_2 \\ r_3 \end{bmatrix} = \begin{bmatrix} 14 & 11 & 13 & 9 \\ 9 & 14 & 11 & 13 \\ 13 & 9 & 14 & 11 \\ 11 & 13 & 9 & 14 \end{bmatrix} \begin{bmatrix} a_0 \\ a_1 \\ a_2 \\ a_3 \end{bmatrix}$$

Fig. 1.12 InvMixColumns of AES decryption

AES operates on a 4×4 column-major order matrix of bytes, termed the state, as shown in Fig. 1.10, where the element $S_{r,c}$ is an 8-bit value that corresponds to the row r and column c of the state. Most AES calculations are done in a special finite field.

AES can be described as follows:

- KeyExpansion—round keys are derived from the key using AES key schedule. AES requires a separate 128-bit round key block for each round plus one more.
- InitialRound

   AddRoundKey—each byte of the state is combined with a block of the round key using bitwise XOR.
- Rounds

   SubBytes—a nonlinear substitution step where each byte is replaced with another according to a lookup table.

   ShiftRows—a transposition step where the last three rows of the state are shifted cyclically a certain number of steps.

   MixColumns—a mixing operation which operates on the columns of the state, combining the four bytes in each column. MixColumns for encryption is defined as in Fig. 1.11, while InvMixColumns for decryption is defined as in Fig. 1.12.

    AddRoundKey
- Final Round (no MixColumns)
    SubBytes
    ShiftRows
    AddRoundKey.

Until May 2009, the only successful published attacks against the full AES were side-channel attacks on some specific implementations. NSA reviewed all the AES finalists, including Rijndael, and stated that all of them were secure enough for US government non-classified data. In June 2003, the US government announced that AES could be used to protect classified information.

The design and strength of all key lengths of the AES algorithm (i.e., 128, 192, and 256) are sufficient to protect classified information up to the SECRET level. TOP SECRET information will require use of either the 192 or 256 key lengths. The implementation of AES in products intended to protect national security systems and/or information must be reviewed and certified by the NSA prior to their acquisition and use.

1.3 Public-Key Encryption

1.3.1 Public-Key Encryption Model

During the early history of encryption, two parties would rely upon a key that they would exchange between themselves by means of a secure method. For example, a face-to-face meeting or an exchange, via a trusted courier, could be used. This key, which both parties kept absolutely secret, could then be used to exchange encrypted messages. A number of significant practical difficulties arise with this approach to distributing keys.

Public-key encryption addresses these drawbacks so that users can communicate securely over a public channel without having to agree upon a shared key beforehand.

The public-key encryption model, as shown in Fig. 1.13, was introduced in 1976 by Whitfield Diffie and Martin Hellman [10] who, influenced by Ralph Merkle's work on public-key distribution, disclosed a method of public-key agreement.

Public-key encryption, also called asymmetric key encryption, is a class of algorithms which require two separate keys, one of which is secret (or private) and one of which is public. Although different, the two parts of this key pair are mathematically linked. The public key is used to encrypt plaintext; whereas the private key is used to decrypt ciphertext. The term "asymmetric" stems from the use of different keys to perform these opposite functions, each the inverse of the other, as contrasted with conventional ("symmetric key") encryption which relies on the same key to perform both.

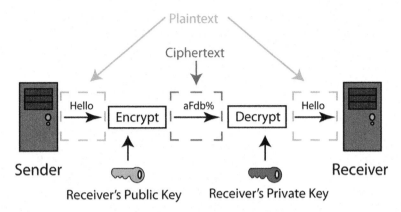

Fig. 1.13 Public-key encryption model

In general, a public-key cryptosystem, associated with a key space (**K**), a plaintext space **M**, and a ciphertext space **C**, consists of three algorithms as follows:

- Key generation algorithm (KG)—given a security parameter k, a public and private key pair (pk, sk) is generated, where $sk \in$ **K**. The public key pk is published to the public, while the private key sk is known to its owner only.
- Encryption algorithm (E)—given a plaintext $m \in$ **M** and a public key pk, a ciphertext c is produced, denoted as $c = E(m, pk)$, where $c \in$ **C**.
- Decryption algorithm (D)—given a ciphertext $c = E(m, pk)$ and the private key sk, the plaintext m is recovered, denoted as $m = D(c, sk)$.

The encryption algorithm E, a map from the plaintext space **M** to the ciphertext space **C**, must be a trapdoor one-way function. For virtually all ciphertexts $c = E(m, pk)$, it must be computationally infeasible to recover the plaintext m from a given pk and c. However, since the legitimate recipient of the message must be able to recover m from c, more is required of the one-way function. Specially, each E must have an inverse D, and this inverse must be easily obtainable given some additional secret information sk. The extra information sk is called a trapdoor of E and the function E itself is called trapdoor one-way function. It is also required that, with a knowledge of sk, $m = D(c, sk)$ be easy to compute for all c in the ciphertext space.

Trapdoor functions are based on mathematical problems which currently admit no efficient solution that are inherent in certain integer factorization, discrete logarithm, and elliptic curve relationships. It is computationally easy for a user to generate their own public- and private key pair and to use them for encryption and decryption. The strength lies in the fact that it is "impossible" (computationally unfeasible) for a properly generated private key to be determined from its corresponding public key. Thus the public key may be published without compromising security, whereas the private key must not be revealed to anyone not authorized to read messages. Public-key algorithms, unlike secret key algorithms, do not require a secure initial exchange of one (or more) secret keys between the parties.

1.3.2   RSA

Diffie and Hellman introduced the great idea of public-key cryptosystem in 1976, but they did not provide a practical public-key cryptosystem. In 1977, the first practicable public-key cryptosystem, RSA [18], was proposed by Ron Rivest, Adi Shamir, and Leonard Adleman and named by their names. In RSA, the encryption key is public and differs from the decryption key which is kept secret, and the security is based on the practical difficulty of factoring the product of two large prime numbers, the factoring problem. Clifford Cocks, an English mathematician, had developed an equivalent system in 1973, but it was not declassified until 1997.

The RSA algorithm involves three algorithms: key generation, encryption, and decryption algorithms as follows.

Key Generation: RSA involves a public key and a private key. The public key can be known by everyone and is used for encrypting messages. Messages encrypted with the public key can only be decrypted in a reasonable amount of time using the private key. The keys for the RSA algorithm are generated in the following way:

1. Choose two distinct prime numbers p and q. For security purposes, the integers p and q should be chosen at random and should be of similar bit-length. Prime integers can be efficiently found using a primality test.
2. Compute

$$n = pq \tag{1.1}$$

   n is used as the modulus for both the public and private keys. Its length, usually expressed in bits, is the key length.
3. Compute

$$\phi(n) = \phi(p)\phi(q) = (p-1)(q-1) \tag{1.2}$$

   where ϕ is Euler's totient function (i.e., the number of positive integers less than n and relatively prime to n).
4. Choose an integer e such that $1 < e < \phi(n)$ and

$$gcd(e, \phi(n)) = 1$$

   In other words, e and $\phi(n)$ are coprime. e is released as the public-key exponent. e having a short bit-length and small Hamming weight results in more efficient encryption, most commonly $2^{16} + 1 = 65,537$. However, much smaller values of e (such as 3) have been shown to be less secure in some settings [4].

5. Determine d as

$$d = e^{-1}(mod\ \phi(n)) \qquad (1.3)$$

that is, d is the multiplicative inverse of $e(mod\ \phi(n))$.

This is more clearly stated as solve for d given $e \cdot d = 1(mod \phi(n))$. This is often computed using the extended Euclidean algorithm. The public key consists of the modulus n and the public (or encryption) exponent e. The private key consists of the modulus n and the private (or decryption) exponent d, which must be kept secret. p, q, and $\phi(n)$ must also be kept secret because they can be used to calculate d.

Encryption: Alice transmits her public key (n, e) to Bob and keeps the private key secret. Bob then wishes to send message M to Alice.

He first turns M into an integer m, such that $0 < m < n$ by using an agreed-upon reversible protocol known as a padding scheme. He then computes the ciphertext c corresponding to

$$c = m^e(mod\ n) \qquad (1.4)$$

This can be done quickly using the method of exponentiation by squaring. Bob then transmits c to Alice.

Decryption: Alice can recover m from c by using her private key exponent d by computing

$$m = c^d(mod\ n) \qquad (1.5)$$

Given m, she can recover the original message M by reversing the padding scheme.

RSA Example: The parameters used here are artificially small, but one can also use OpenSSL to generate and examine a real key pair.

1. Choose two distinct prime numbers, such as

$$p = 61, q = 53$$

2. Compute $n = pq$ giving

$$n = 61 \times 53 = 3233$$

3. Compute the totient of the product as $\phi(n) = (p-1)(q-1)$, giving

$$\phi(3233) = (61 - 1) \times (53 - 1) = 3120$$

4. Choose any number $1 < e < 3120$ that is coprime to 3120. Choosing a prime number for e leaves us only to check that e is not a divisor of 3120. Let

$$e = 17$$

5. Compute d, the modular multiplicative inverse of $e(mod\phi(n))$ yielding

$$d = 2753$$

The public key is ($n = 3233, e = 17$). For a padded plaintext message m, the encryption function is

$$c = m^{17}(mod\ 3233).$$

The private key is ($n = 3233, d = 2753$). For an encrypted ciphertext c, the decryption function is

$$m = c^{2753}(mod\ 3233).$$

For instance, in order to encrypt

$$m = 65$$

we calculate

$$c = 65^{17} = 2790(mod\ 3233).$$

To decrypt $c = 2790$, we calculate

$$m = 2790^{2753}(mod\ 3233) = 65.$$

RSA Security: There are a number of attacks against plain RSA as described below:

1. When encrypting with low encryption exponents (e.g., $e = 3$) and small values of the m (i.e., $m < n^{1/e}$) the result of m^e is strictly less than the modulus n. In this case, ciphertexts can be easily decrypted by taking the e-th root of the ciphertext over the integers.
2. If the same clear text message is sent to e or more recipients in an encrypted way, and the receivers share the same exponent e, but different p, q, and therefore n, then it is easy to decrypt the original clear text message via the Chinese remainder theorem. Johan Hastad [11] noticed that this attack is possible even if the cleartexts are not equal, but the attacker knows a linear relation between them. This attack was later improved by Don Coppersmith [6].

3. Because RSA encryption is a deterministic encryption algorithm (i.e., has no random component), an attacker can successfully launch a chosen-plaintext attack against the cryptosystem, by encrypting likely plaintexts under the public key and test if they are equal to the ciphertext. A cryptosystem is called semantically secure if an attacker cannot distinguish two encryptions from each other even if the attacker knows (or has chosen) the corresponding plaintexts. As described above, RSA without padding is not semantically secure.

4. RSA has the property that the product of two ciphertexts is equal to the encryption of the product of the respective plaintexts. That is, $m_1^e m_2^e = (m_1 m_2)^e \pmod{n}$. Because of this multiplicative property, a chosen-ciphertext attack is possible, e.g., an attacker, who wants to know the decryption of a ciphertext $c = m^e \pmod{n}$ may ask the holder of the private key to decrypt an unsuspicious-looking ciphertext $c' = c r^e \pmod{n}$ for some value r chosen by the attacker. Because of the multiplicative property c' is the encryption of $m r \pmod{n}$. Hence, if the attacker is successful with the attack, he or she will learn $m r \pmod{n}$ from which he or she can derive the message m by multiplying $m r$ with the modular inverse of r modulo n.

To avoid these problems, practical RSA implementations typically embed some form of structured, randomized padding into the value m before encrypting it. This padding ensures that m does not fall into the range of insecure plaintexts and that a given message, once padded, will encrypt to one of a large number of different possible ciphertexts.

The security of RSA is based on two mathematical problems: the problem of factoring large numbers and the RSA problem. It is easy to multiply two large prime numbers, but no algorithm is known that is able to factorize a large number efficiently. The RSA problem is defined as the task of taking eth roots modulo a composite n: recovering a value m such that $c = m^e \pmod{n}$, where (n, e) is an RSA public key and c is an RSA ciphertext. Currently the most promising approach to solving the RSA problem is to factor the modulus n. With the ability to recover prime factors, an attacker can compute the secret exponent d from a public key (n, e), then decrypt c using the standard procedure. To accomplish this, an attacker factors n into p and q and computes $(p - 1)(q - 1)$ which allows the determination of d from e.

However, it is not proved that RSA is as secure as factoring. It can be shown that if an attacker is able to generate a private key from a public key, he or she is also able to factorize large numbers. But until today nobody was able to prove that an attacker who is able to decrypt messages is also able to factorize large numbers. So it is unknown if the complexity of the RSA problem is the same as the complexity of factoring.

1.3.3 Rabin Public-Key Encryption

In 1979, two years after the publication of RSA, Michael O. Rabin [17] proposed the
Rabin public-key algorithm, which has the advantage of being provably as secure as
factoring.

As with all asymmetric cryptosystems, the Rabin system uses both a public and
a private key. The public key is necessary for later encryption and can be published,
while the private key must be possessed only by the recipient of the message.

Key Generation: The precise key generation process is as follows:

- Choose two large distinct primes p and q. One may choose $p = q = 3(mod\ 4)$
  to simplify the computation of square roots modulo p and q. But the scheme
  works with any primes.
- Let $n = p \cdot q$. Then n is the public key. The primes p and q are the private key.
  To encrypt a message only the public key n is needed. To decrypt a ciphertext the
  factors p and q of n are necessary.

As a (non-real-world) example, if $p = 7$ and $q = 11$, then $n = 77$. The public
key, 77, would be released, and the message encoded using this key. And, in order to
decode the message, the private keys, 7 and 11, would have to be known (of course,
this would be a poor choice of keys, as the factorization of 77 is trivial; in reality
much larger numbers would be used).

Encryption: For the encryption, only the public key n is used, thus producing a
ciphertext out of the plaintext. The process is as follows:

Let $P = \{0,\ldots,n-1\}$ be the plaintext space (consisting of numbers) and $m \in P$
be the plaintext. Now the ciphertext c is determined by

$$c = m^2 (mod\ n) \tag{1.6}$$

That is, c is the quadratic remainder of the square of the plaintext, modulo the
public key n.

In the simple example, $P = \{0,\ldots,76\}$ is the plaintext space. We will take $m =$
20 as the plaintext. The ciphertext is thus $c = m^2(mod\ n) = 400(mod\ 77) = 15$.
For exactly four different values of m, the ciphertext 15 is produced, i.e., for $m \in$
$\{13, 20, 57, 64\}$. This is true for most ciphertexts produced by the Rabin algorithm,
i.e., it is a four-to-one function.

Decryption: To decode the ciphertext, the private keys are necessary. The process
is as follows:

If c and n are known, the plaintext is then $m \in \{0,\ldots,n-1\}$ with $m^2 =$
$c(mod\ n)$. For a composite n (that is, like the Rabin algorithm's $n = p \cdot q$) there is
no efficient method known for the finding of m. If, however n is prime (as are p and
q in the Rabin algorithm), the Chinese remainder theorem can be applied to solve
for m.

Thus the square roots

$$m_p = \sqrt{c}(mod\ p) \tag{1.7}$$

$$m_q = \sqrt{c}(mod\ q) \tag{1.8}$$

must be calculated.

When $p = q = 3(mod\ 4)$, we can compute square roots by

$$m_p = c^{\frac{1}{4}(p+1)}(mod\ p) \tag{1.9}$$

$$m_q = c^{\frac{1}{4}(q+1)}(mod\ q) \tag{1.10}$$

In the example we get $m_p = 1$ and $m_q = 9$.

By applying the extended Euclidean algorithm, we wish to find y_p and y_q such that $y_p \cdot p + y_q \cdot q = 1$. In the example, we have $y_p = -3$ and $y_q = 2$.

Now, by invocation of the Chinese remainder theorem, the four square roots $+r, -r, +s$, and $-s$ of $c + n\mathbb{Z} \in \mathbb{Z}/n\mathbb{Z}$ are calculated ($\mathbb{Z}/n\mathbb{Z}$ here stands for the ring of congruence classes modulo n). The four square roots are in the set $\{0, \ldots, n-1\}$:

$$r = (y_p \cdot p \cdot m_q + y_q \cdot q \cdot m_p)\ mod\ n \tag{1.11}$$

$$-r = n - r \tag{1.12}$$

$$s = (y_p \cdot p \cdot m_q - y_q \cdot q \cdot m_p)\ mod\ n \tag{1.13}$$

$$-s = n - s \tag{1.14}$$

One of these square roots $mod\ n$ is the original plaintext m. In the example, $m \in \{64, \mathbf{20}, 13, 57\}$.

Security: Rabin pointed out in his paper that if someone is able to compute both, r and s, then he is also able to find the factorization of n because either $\gcd(|r - s|, n) = p$ or $\gcd(|r - s|, n) = q$, where gcd means greatest common divisor. Since the greatest common divisor can be calculated efficiently, you are able to find the factorization of n efficiently if you know r and s. In the our example (picking 57 and 13 as r and s):

$$\gcd(57 - 13, 77) = \gcd(44, 77) = 11 = q$$

Rabin scheme has, however, a downside: Every decryption operation produces four possible outputs and thus is not suitable for practical applications. Williams [22] suggested a change that avoids these ambiguities. This is called the Rabin–Williams algorithm.

1.3.4   Public-Key Cryptography Standards

The public-key cryptography standards (PKCS) are a set of standard protocols for making possible secure information exchange on the Internet using a public-key infrastructure (PKI). The standards include RSA encryption, password-based encryption, extended certificate syntax, and cryptographic message syntax for S/MIME, RSA's proposed standard for secure e-mail. The standards were developed by RSA laboratories in cooperation with a consortium that included Apple, Microsoft, DEC, Lotus, Sun, and MIT.

Public-key cryptography standards (PKCS) #1 provides the basic definitions of and recommendations for implementing the RSA algorithm for public-key cryptography. It defines the mathematical properties of public and private keys, primitive operations for encryption and signatures, secure cryptographic schemes, and related ASN.1 syntax representations. The current version, 2.1, was published in June 2002 and was also republished as RFC 3447 in February 2003.

Standards such as PKCS#1 have been carefully designed to securely pad messages prior to RSA encryption. Because these schemes pad the plaintext m with some number of additional bits, the size of the un-padded message must be somewhat smaller. RSA padding schemes must be carefully designed so as to prevent sophisticated attacks which may be facilitated by a predictable message structure. Early versions of the PKCS#1 standard (up to version 1.5) used a construction that appears to make RSA semantically secure. However, at Eurocrypt 2000, Coron et al. [7] showed that for some types of messages, this padding does not provide a high enough level of security. Furthermore, at Crypto 1998, Bleichenbacher [3] showed that this version is vulnerable to a practical adaptive chosen-ciphertext attack. Later versions of the standard include optimal asymmetric encryption padding (OAEP) [1], which prevents these attacks. As such, OAEP should be used in any new application, and PKCS#1 v1.5 padding should be replaced wherever possible.

Optimal asymmetric encryption padding (OAEP) is a padding scheme often used together with RSA encryption. OAEP was introduced by Bellare and Rogaway [1] and subsequently standardized in PKCS #1v2 and RFC 2437.

The OAEP algorithm is a form of Feistel network which uses a pair of random oracles G and H to process the plaintext prior to asymmetric encryption. When combined with any secure trapdoor one-way permutation, this processing is proved in the random oracle model to result in a combined scheme which is semantically secure under chosen-plaintext attack (IND-CPA). When implemented with certain trapdoor permutations (e.g., RSA), OAEP is also proved secure against chosen-ciphertext attack. OAEP can be used to build an all-or-nothing transform.

The OAEP algorithm can be depicted in a diagram as shown in Fig. 1.14. In the diagram,

- n is the number of bits in the RSA modulus.
- k_0 and k_1 are integers fixed by the protocol.

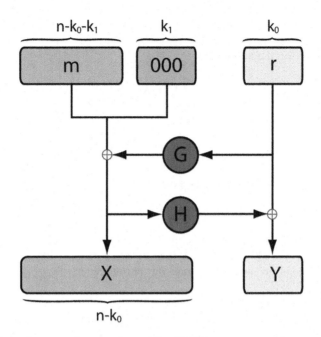

Fig. 1.14 Optimal asymmetric encryption padding (OAEP)

- m is the plaintext message, an $(n - k_0 - k_1)$-bit string
- G and H are typically some cryptographic hash functions fixed by the protocol.

To encode,

Step 1.   Pad messages with k_1 zeros to be $n - k_0$ bits in length.
Step 2.   Generate a random number r with k_0-bit string
Step 3.   Expand the k_0 bits of r to $n - k_0$ bits with G.
Step 4.   Let

$$X = m00..0 \oplus G(r) \tag{1.15}$$

Step 5.   Reduce the $n - k_0$ bits of X to k_0 bits with H.
Step 6.   Let

$$Y = r \oplus H(X) \tag{1.16}$$

Step 7.   The output is $X \| Y$ where X is shown in the diagram as the leftmost block and Y as the rightmost block.

To decode,

Step 1'.   Recover the random string as

$$r = Y \oplus H(X) \tag{1.17}$$

Step 2'.   Recover the message as

$$m00..0 = X \oplus G(r) \tag{1.18}$$

The "all-or-nothing" security is from the fact that to recover m, you must recover the entire X and the entire Y; X is required to recover r from Y, and r is required to recover m from X. Since any changed bit of a cryptographic hash completely changes the result, the entire X and the entire Y must both be completely recovered.

OAEP satisfies the following two goals:

- Add an element of randomness which can be used to convert a deterministic encryption scheme (e.g., traditional RSA) into a probabilistic scheme.
- Prevent partial decryption of ciphertexts (or other information leakage) by ensuring that an adversary cannot recover any portion of the plaintext without being able to invert the trapdoor one-way permutation.

References

1. M. Bellare, P. Rogaway, Optimal asymmetric encryption—how to encrypt with RSA, in *Proceedings of Eurocrypt'94*, 1994
2. E. Biham, A. Shamir, Differential cryptanalysis of DES-like cryptosystems. J. Cryptol. **4**(1), 3–72 (1991)
3. D. Bleichenbacher, Chosen ciphertext attacks against protocols based on the RSA encryption standard PKCS #1, in *Proceedings of the CRYPTO'98*, 1998, pp. 1–12
4. D. Boneh, Twenty years of attacks on the RSA cryptosystem. Not. Am. Math. Soc. **46**(2), 203–213 (1999)
5. A. Bruen, M. Forcinito, *Cryptography, Information Theory, and Error-Correction: A Handbook for the 21st Century* (Wiley, Newyork, 2011), p. 21
6. D. Coppersmith, Small solutions to polynomial equations, and low exponent RSA vulnerabilities. J. Cryptol. **10**(4), 233–260 (1997)
7. J. Coron, D. Naccache, Security analysis of the Gennaro-Halevi-Rabin signature scheme, in *Proceedings of EUROCRYPT'00*, 2000, pp. 91–101
8. J. Daemen, V. Rijmen, *AES Proposal: Rijndael* (National Institute of Standards and Technology, 2013), p.1,
9. D. Davies, S. Murphy, Pairs and triplets of DES S-boxes. J. Cryptol. **8**(1), 1–25 (2007)
10. W. Diffie, M. Hellman, New directions in cryptography. IEEE Trans. Inf. Theory **22**(6), 644–654 (1976)
11. J. Hastad, On using RSA with low exponent in a public key network, in *Proceedings of CRYPTO'85*, 1986, pp. 403–408
12. D. Kahn, *The Codebreakers: The Story of Secret Writing* (Rev Sub. Scribner, NewYork, 1996)
13. T. Kelly. The myth of the skytale. Cryptologia **22**(3), 244–260 (1998)
14. D. Luciano, G. Prichett, Cryptology: from Caesar ciphers to public-key cryptosystems. Coll. Math. J. **18**(1), 2–17 (1987)
15. M. Matsui, Linear cryptanalysis method for DES cipher, in *Proceedings of EUROCRYPT'93*, 1994, pp. 386–397

16. D. Newton, *Encyclopedia of Cryptography. Instructional Horizons, Inc* (Santa Barbara, California, 1997), p. 6
17. M. Rabin, *Digitalized Signatures and Public-Key Functions as Intractable as Factorization* (MIT Laboratory for Computer Science, Cambridge, 1979)
18. R. Rivest, A. Shamir, L. Adleman, A method for obtaining digital signatures and public-key cryptosystems. Commun. ACM **21**(2), 120–126 (1978)
19. B. Schneier, *Applied Cryptography, Protocols, Algorithms, and Source Code in C*, 2nd edn. (Wiley, New York, 1996), p. 267
20. C. Shannon, Communication theory of secrecy systems. Bell Syst. Techn. J. **28**(4), 656–715 (1949)
21. M. Urban, The lockade of Ciudad Rodrigo, June to November 1811—The great cipher. *In The Man Who Broke Napoleon's Codes* (Harper Perennial, New York, 2003)
22. H. Williams, A modification of the RSA public-key encryption procedure. IEEE Trans. Inform. Theor. **26**(6), 726–729 (1980)

Chapter 2
Homomorphic Encryption

Abstract Homomorphic encryption is a form of encryption which allows specific types of computations to be carried out on ciphertexts and generate an encrypted result which, when decrypted, matches the result of operations performed on the plaintexts. This is a desirable feature in modern communication system architectures. RSA is the first public-key encryption scheme with a homomorphic property. However, for security, RSA has to pad a message with random bits before encryption to achieve semantic security. The padding results in RSA losing the homomorphic property. To avoid padding messages, many public-key encryption schemes with various homomorphic properties have been proposed in last three decades. In this chapter, we introduce basic homomorphic encryption techniques. It begins with a formal definition of homomorphic encryption, followed by some well-known homomorphic encryption schemes.

2.1 Homomorphic Encryption Definition

In abstract algebra, a homomorphism is a structure-preserving map between two algebraic structures, such as groups.

A group is a set, G, together with an operation $\circ$ (called the group law of G) that combines any two elements a and b to form another element, denoted $a \circ b$. To qualify as a group, the set and operation, $(G, \circ)$, must satisfy four requirements known as the group axioms:

- Closure: For all a, b in G, the result of the operation, $a \circ b$, is also in G.
- Associativity: For all a, b, and c in G, $(a \circ b) \circ c = a \circ (b \circ c)$.
- Identity element: There exists an element e in G, such that for every element a in G, the equality $e \circ a = a \circ e = a$ holds. Such an element is unique, and thus one speaks of the identity element.
- Inverse element: For each a in G, there exists an element b in G such that $a \circ b = b \circ a = e$, where e is the identity element.

The identity element of a group G is often written as 1.

The result of an operation may depend on the order of the operands. In other words, the result of combining element a with element b need not yield the same result as combining element b with element a; the equation $a \circ b = b \circ a$ may not

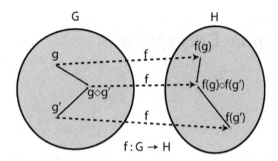

Fig. 2.1 Group Homomorphism

always be true. This equation always holds in the group of integers under addition, because $a + b = b + a$ for any two integers (commutativity of addition). Groups for which the commutativity equation $a \circ b = b \circ a$ always holds are called abelian groups.

Given two groups $(G, \diamond)$ and $(H, \circ)$, a group homomorphism from $(G, \diamond)$ to $(H, \circ)$ is a function $f : G \to H$ such that for all g and g' in G it holds that

$$f(g \diamond g') = f(g) \circ f(g') \tag{2.1}$$

Group homomorphism can be illustrated as in Fig. 2.1.

Let (P, C, K, E, D) be an encryption scheme, where P, C are the plaintext and ciphertext spaces, K is the key space, and E, D are the encryption and decryption algorithms. Assume that the plaintexts forms a group $(P, \diamond)$ and the ciphertexts forms a group $(C, \circ)$, then the encryption algorithm E is a map from the group P to the group C, i.e., $E_k : P \to C$, where $k \in K$ is either a secret key (in a secret key cryptosystem) or a public key (in a public-key cryptosystem).

For all a and b in P and k in K, if

$$E_k(a) \circ E_k(b) = E_k(a \diamond b) \tag{2.2}$$

the encryption scheme is **homomorphic**.

In an unpadded RSA [18], assume that the public key $pk = (n, e)$, the plaintexts form a group $(P, \cdot)$, and the ciphertexts form a group $(C, \cdot)$, where $\cdot$ is the modular multiplication. For any two plaintexts m_1, m_2 in P, it holds that

$$E(m_1, pk) \cdot E(m_2, pk) = m_1^e \cdot m_2^e (mod\ n)$$
$$= (m_1 \cdot m_2)^e (mod\ n)$$
$$= E(m_1 \cdot m_2, pk)$$

Therefore, the unpadded RSA has the homomorphic property. Unfortunately, the unpadded RSA is insecure.

2.2 Goldwasser–Micali Encryption Scheme

The Goldwasser–Micali (GM) encryption scheme [7] is a public-key encryption algorithm developed by Shafi Goldwasser and Silvio Micali in 1982. GM has the distinction of being the first probabilistic public-key encryption scheme which is provably secure under standard cryptographic assumptions. However, it is not an efficient cryptosystem, as ciphertexts may be several hundred times larger than the initial plaintext. To prove the security properties of the cryptosystem, Goldwasser and Micali proposed the widely used definition of semantic security.

GM consists of three algorithms: a probabilistic key generation algorithm which produces a public and a private key, a probabilistic encryption algorithm, and a deterministic decryption algorithm.

The scheme relies on deciding whether a given value x is a square mod N, given the factorization (p, q) of N. This can be accomplished using the following procedure:

Compute

$$x_p = x(mod\ p) \tag{2.3}$$

$$x_q = x(mod\ q) \tag{2.4}$$

If

$$x_p^{(p-1)/2} = 1(mod\ p) \tag{2.5}$$

$$x_q^{(q-1)/2} = 1(mod\ q) \tag{2.6}$$

then x is a quadratic residue mod N.

Key Generation: The modulus used in GM encryption is generated in the same manner as in the RSA cryptosystem.

Alice generates two distinct large prime numbers p and q, such that $p = q = 3(mod\ 4)$, randomly and independently of each other. Alice computes $N = pq$. She then finds some non-residue a such that

$$a_p^{(p-1)/2} = -1(mod\ p), a_q^{(q-1)/2} = -1(mod\ q)$$

The public key consists of (a, N). The secret key is the factorization (p, q).

Encryption: Suppose Bob wishes to send a message m to Alice. Bob first encodes m as a string of bits $(m_1, \cdots, m_n)$.

For every bit m_i, Bob generates a random value b_i from the group of units modulo N, or $gcd(b_i, N) = 1$. He outputs the value

$$c_i = b_i^2 \cdot a^{m_i} (mod\ N) \tag{2.7}$$

Bob sends the ciphertext $(c_1, c_2, \cdots, c_n)$ to Alice.

Decryption: Alice receives $(c_1, c_2, \cdots, c_n)$. She can recover m using the following procedure:

For each i, using the prime factorization (p, q), Alice determines whether the value c_i is a quadratic residue; if so, $m_i = 0$, otherwise $m_i = 1$. Alice outputs the message $m = (m_1, \cdots, m_n)$.

GM Example: We choose small parameters in this example. In key generation, we let

$$p = 7, q = 11$$

where $p = q = 3 \pmod 4$. So

$$N = pq = 77$$

Take

$$a = 6$$

where

$$6^{(7-1)/2} = -1 \pmod 7, 6^{(11-1)/2} = -1 \pmod{11}$$

The public key is $(6, 77)$ and the private key is $(7, 11)$.
To encrypt 3-bit message $m_1 m_2 m_3 = 101$. Choose

$$b_1 = 2, b_2 = 3, b_3 = 5$$

and compute

$$c_1 = 2^2 \cdot 6^1 = 24 \pmod{77}$$

$$c_2 = 3^2 \cdot 6^0 = 9 \pmod{77}$$

$$c_3 = 5^2 \cdot 6^1 = 73 \pmod{77}$$

The ciphertext is $(24, 9, 73)$.
To decrypt the ciphertext, compute

$$24^{(7-1)/2} = -1 \pmod 7$$

$$9^{(7-1)/2} = 1 \pmod 7, 9^{(11-1)/2} = 1 \pmod{11}$$

$$73^{(7-1)/2} = -1 \pmod 7$$

This shows that 24 and 73 are non-quadratic residue and 9 is quadratic residue, and thus outputs the plaintext 101.

Homomorphic Property: The GM encryption scheme has a homomorphic property, in the sense that if c_0, c_1 are the encryptions of bits m_0, m_1, then $c_0 c_1 (mod\ N)$ will be an encryption of $m_0 \oplus m_1$, where $\oplus$ denotes addition modulo 2 (i.e., exclusive-OR).

Assume that

$$c_0 = b_0^2 \cdot a^{m_0} (mod\ N), c_1 = b_1^2 \cdot a^{m_1} (mod\ N)$$

we have

$$c_0 \cdot c_1 = (b_0^2 \cdot a^{m_0}) \cdot (b_1^2 \cdot a^{m_1})(mod\ N)$$
$$= (b_0 b_1)^2 \cdot a^{m_0 + m_1} (mod\ N)$$

When $m_0 + m_1$ is either 0 or 1, we have $m_0 + m_1 = m_0 \oplus m_1$. When $m_0 = m_1 = 1$, $m_0 + m_1 = 2$ and $c_0 c_1 (mod\ N)$ is a quadratic residue and thus it is an encryption of 0. In this case, we have $m_0 \oplus m_1 = 1 \oplus 1 = 0$ as well.

Security: The GM encryption scheme is a probabilistic encryption [8]. Probabilistic encryption refers to the use of randomness in an encryption algorithm, so that when encrypting the same message several times it will, in general, yield different ciphertexts. The term "probabilistic encryption" is typically used in reference to public-key encryption algorithms; however, various secret key encryption algorithms achieve a similar property (e.g., block ciphers when used in a chaining mode such as CBC). To be semantically secure, that is, to hide even partial information about the plaintext, an encryption algorithm must be probabilistic.

Probabilistic encryption is particularly important when using public-key encryption. Suppose that the adversary observes a ciphertext and suspects that the plaintext is either "YES" or "NO." When a deterministic encryption algorithm is used, the adversary can simply try encrypting each of his or her guesses under the recipient's public key and compare each result to the target ciphertext. To combat this attack, public-key encryption schemes must incorporate an element of randomness, ensuring that each plaintext maps into one of a large number of possible ciphertexts.

An intuitive approach to converting a deterministic encryption scheme into a probabilistic one is to simply pad the plaintext with a random string before encrypting with the deterministic algorithm, such as padding RSA. Conversely, decryption involves applying a deterministic algorithm and ignoring the random padding. However, early schemes which applied this naive approach were broken due to limitations in some deterministic encryption schemes. Techniques such as OAEP integrate random padding in a manner that is secure using any trapdoor permutation.

The GM encryption scheme is semantically secure [8]. Semantic security is commonly defined by the following game:

- *Initialize*: The challenger runs the key generation algorithm, gives the public key pk to a probabilistic polynomial time-bounded (PPT) adversary, but keeps the private key sk to itself.
- *Phase* 1: The adversary adaptively asks a number of different encryption queries $C_i = \mathsf{E}(m_i, pk)$ for m_i, where $i = 1, 2, \cdots, n$.
- *Challenge*: Once the adversary decides that Phase 1 is over, it outputs a pair of equal length plaintexts (M_0, M_1) on which it wishes to be challenged. The challenger picks a random bit $b \in \{0, 1\}$ and sends $C = \mathsf{E}(M_b, pk)$ as the challenge to the adversary.
- *Phase* 2: The adversary issues more encryption queries adaptively as in Phase 1.
- *Guess*: Finally, the adversary outputs a guess $b' \in \{0, 1\}$ and wins the game if $b' = b$.

The public-key encryption cryptosystem is semantically secure under chosen-plaintext attack if the adversary cannot determine which of the two messages was chosen by the challenger, with probability significantly greater than 1/2 (the success rate of random guessing).

The GM encryption scheme is semantically secure based on the assumed intractability of the quadratic residuosity problem modulo a composite $N = pq$ where p, q are large primes. This assumption states that given (a, N) it is difficult to determine whether a is a quadratic residue modulo N (i.e., $a = b^2 (mod\ N)$ for some b). The quadratic residue problem is easily solved given the factorization of N. The GM encryption scheme leverages this asymmetry by encrypting individual plaintext bits as either random quadratic residues or non-residues modulo N. Recipients use the factorization of N as a secret key and decrypt the message by testing the quadratic residuosity of the received ciphertext values.

Because the GM encryption scheme produces a value of size approximately $|N|$ to encrypt every single bit of a plaintext, GM encryption results in substantial ciphertext expansion. To prevent factorization attacks, it is recommended that $|N|$ be several hundred bits or more. Thus, the scheme serves mainly as a proof of concept, and more efficient provably secure schemes such as ElGamal encryption scheme have been developed since.

2.3 ElGamal Encryption Scheme

The ElGamal encryption scheme [4] is a public-key encryption algorithm based on the Diffie–Hellman key exchange. It was invented by Taher Elgamal in 1985. The ElGamal encryption scheme is used in the free GNU Privacy Guard software, recent versions of PGP, and other cryptosystems. The ElGamal encryption scheme can be defined over any cyclic group G. Its security depends upon the difficulty of a certain problem in G related to computing discrete logarithms.

The ElGamal encryption scheme consists of three components: the key generation, the encryption algorithm, and the decryption algorithm.

Key Generation: The key generator works as follows:

Alice generates an efficient description of a cyclic group G, of order q, with generator g.

Alice chooses a random $x \in \{1, \ldots, q - 1\}$.

Alice computes

$$y = g^x \tag{2.8}$$

Alice publishes y along with the description of G, q, g, as her public key. Alice retains x, as her private key which must be kept secret.

Encryption: The encryption algorithm works as follows:

To encrypt a message m, to Alice under her public key (G, q, g, y), Bob chooses a random $r \in \{1, \ldots, q - 1\}$, then computes

$$c_1 = g^r \tag{2.9}$$

Bob computes the shared secret

$$s = y^r \tag{2.10}$$

Bob converts his secret message m, into an element $m' \in G$.

Bob computes

$$c_2 = m' \cdot s \tag{2.11}$$

Bob sends the ciphertext $(c_1, c_2) = (g^r, m' \cdot y^r)$ to Alice.

Note that one can easily find y^r, if one knows m'. Therefore, a new r, is generated for every message to improve security. For this reason, r, is also called an ephemeral key.

Decryption: The decryption algorithm works as follows:

To decrypt a ciphertext (c_1, c_2), with her private key x, Alice computes the shared secret

$$t = c_1^x \tag{2.12}$$

and then computes

$$m' = c_2 \cdot t^{-1} \tag{2.13}$$

which she then converts back into the plaintext message m, where t^{-1} is the inverse of t in the group G (e.g., modular multiplicative inverse if G is a subgroup of a multiplicative group of integers modulo n).

The decryption algorithm produces the intended message, since

$$\begin{aligned} c_2 \cdot t^{-1} &= (m' \cdot s) \cdot c_1^{-x} \\ &= m' \cdot y^r \cdot g^{-xr} \\ &= m' \cdot g^{xr} \cdot g^{-xr} \\ &= m' \end{aligned}$$

The ElGamal encryption scheme is probabilistic, meaning that a single plaintext can be encrypted to many possible ciphertexts, with the consequence that a general ElGamal encryption produces a 2:1 expansion in size from plaintext to ciphertext.

Encryption under ElGamal requires two exponentiations; however, these exponentiations are independent of the message and can be computed ahead of time if need be. Decryption only requires one exponentiation.

The division by t can be avoided by using an alternative method for decryption. To decrypt a ciphertext (c_1, c_2), with Alice's private key x, Alice computes $t' = c_1^{q-x} = g^{(q-x)r}$. t' is the inverse of t. This is a consequence of Lagrange's theorem, because

$$t \cdot t' = g^{xr} \cdot g^{(q-x)r} = (g^q)^r = 1^r = 1$$

where 1 is the identity element of G.

Alice then computes $m' = c_2 \cdot t'$, by which she then converts back into the plaintext message m. The decryption algorithm produces the intended message, since

$$c_2 \cdot t' = m' \cdot s \cdot t' = m' \cdot y^r \cdot t' = m' \cdot g^{xr} \cdot t' = m' \cdot (g^r)^x \cdot t' = m' \cdot c_1^x \cdot t' = m' \cdot t \cdot t' = m'$$

ElGamal Example: An example of the ElGamal encryption with small parameters is given as follows:

At first, Alice generates a prime modulo p and a group generator g which is between 1 and $p - 1$:

$$p = 2879$$

$$g = 2585$$

Alice selects a random number (x) which will be her private key:

$$x = 47$$

She then calculates

$$y = g^x = 2585^{47} = 2826 \pmod{2879}$$

Alice's public key is now (p, g, y) and sends them to Bob. The private key x is known to Alice only.

Bob then creates a message

$$m = 77$$

and then selects a random value

$$r = 65$$

and calculates the ciphertext (c_1, c_2) where

$$c_1 = g^r = 2585^{65} = 319 (mod\ 2879)$$

$$c_2 = m \cdot y^r = 77 \cdot 2826^{65} = 472 (mod\ 2879)$$

Alice can decrypt the ciphertext:

$$c_2 / c_1^x = 472 / 319^{47} = 77 (mod\ 2879).$$

Homomorphic Property: ElGamal encryption scheme has a homomorphic property. Given two encryptions

$$(c_{11}, c_{12}) = (g^{r_1}, m_1 y^{r_1}), (c_{21}, c_{22}) = (g^{r_2}, m_2 y^{r_2})$$

where r_1, r_2 are randomly chosen from $\{1, 2, \cdots, q - 1\}$ and $m_1, m_2 \in G$, one can compute

$$(c_{11}, c_{12})(c_{21}, c_{22}) = (c_{11} c_{21}, c_{12} c_{22})$$
$$= (g^{r_1} g^{r_2}, (m_1 y^{r_1})(m_2 y^{r_2}))$$
$$= (g^{r_1 + r_2}, (m_1 m_2) y^{r_1 + r_2})$$

The resulted ciphertext is an encryption of $m_1 m_2$.

ElGamal Security: The security of the ElGamal scheme depends on the properties of the underlying group G as well as any padding scheme used on the messages.

If the computational Diffie–Hellman assumption (CDH) holds in the underlying cyclic group G, then the ElGamal encryption function is one way. The CDH is the assumption that a certain computational problem within a cyclic group G is hard. Consider a cyclic group G of order q, the CDH assumption states that, given (g, g^a, g^b) for a randomly chosen generator g and random $a, b \in \{0, \cdots, q - 1\}$, it is computationally intractable to compute the value g^{ab}.

If the decisional Diffie–Hellman assumption (DDH) holds in G, then ElGamal achieves semantic security. Semantic security is not implied by the CDH alone. The DDH is a computational hardness assumption about a certain problem involving discrete logarithms in cyclic groups. Consider a (multiplicative) cyclic group G of order q, and with generator g. The DDH assumption states that, given g^a and g^b for uniformly and independently chosen $a, b \in \mathbb{Z}_q$, the value g^{ab} "looks like" a random element in G. This intuitive notion is formally stated by saying that the following two probability distributions are computationally indistinguishable:

- (g^a, g^b, g^{ab}), where a and b are randomly and independently chosen from $\mathbb{Z}_q$;
- (g^a, g^b, g^c), where a, b, c are randomly and independently chosen from $\mathbb{Z}_q$.

ElGamal encryption is unconditionally malleable and therefore is not secure under chosen-ciphertext attack. For example, given an encryption (c_1, c_2) of some (possibly unknown) message m, one can easily construct a valid encryption $(c_1, 2c_2)$ of the message $2m$.

To achieve chosen-ciphertext security, the scheme must be further modified, or an appropriate padding scheme must be used. Depending on the modification, the DDH assumption may or may not be necessary.

Other schemes related to ElGamal which achieve security against chosen-ciphertext attacks have also been proposed. The Cramer–Shoup cryptosystem [3] is secure under chosen-ciphertext attack assuming DDH holds for G. Its proof does not use the random oracle model. Another proposed scheme is DHAES [1], whose proof requires an assumption that is weaker than the DDH assumption.

The ElGamal encryption scheme is usually used in a hybrid cryptosystem, i.e., the message itself is encrypted using a symmetric cryptosystem and ElGamal is then used to encrypt the key used for the symmetric cryptosystem. This is because asymmetric cryptosystems like ElGamal are usually slower than symmetric ones for the same level of security, so it is faster to encrypt the symmetric key (which most of the time is quite small if compared to the size of the message) with ElGamal and the message (which can be arbitrarily large) with a symmetric cryptosystem.

2.4 Paillier Encryption Scheme

The Paillier encryption scheme [11], named after and invented by Pascal Paillier in 1999, is a probabilistic public-key algorithm. The problem of computing nth residue classes is believed to be computationally difficult. The decisional composite residuosity assumption is the intractability hypothesis upon which this cryptosystem is based.

The Paillier encryption scheme is composed of key generation, encryption, and decryption algorithms as follows:

Key Generation: Choose two large prime numbers p and q randomly and independently of each other, such that

$$\gcd(pq, (p-1)(q-1)) = 1$$

This property is assured if both primes are of equal length.

Compute

$$n = pq, \lambda = lcm(p - 1, q - 1)$$

where lcm stands for the least common multiple.

Select random integer g where $g \in \mathbb{Z}_{n^2}^*$.

Ensure n divides the order of g by checking the existence of the following modular multiplicative inverse:

$$\mu = (L(g^\lambda (mod \; n^2)))^{-1}(mod \; n) \tag{2.14}$$

where function L is defined as

$$L(u) = \frac{u - 1}{n} \tag{2.15}$$

Note that the notation a/b does not denote the modular multiplication of a times the modular multiplicative inverse of b, but rather the quotient of a divided by b.

Finally, the public (encryption) key is (n, g) and the private (decryption) key is (λ, μ).

If using p, q of equivalent length, a simpler variant of the above key generation steps would be to set

$$g = n + 1, \lambda = \varphi(n), \mu = \varphi(n)^{-1}(mod \; n)$$

where $\varphi(n) = (p - 1)(q - 1)$.

Encryption: Let m be a message to be encrypted where $m \in \mathbb{Z}_n$.

Select random r where $r \in \mathbb{Z}_n^*$

Compute ciphertext as

$$c = g^m \cdot r^n (mod \; n^2) \tag{2.16}$$

Decryption: Let c be the ciphertext to decrypt, where $c \in \mathbb{Z}_{n^2}^*$

Compute the plaintext message as:

$$m = L(c^\lambda (mod \; n^2)) \cdot \mu(mod \; n) \tag{2.17}$$

As the original paper points out, decryption is "essentially one exponentiation modulo n^2."

The Paillier encryption scheme exploits the fact that certain discrete logarithms can be computed easily. For example, by binomial theorem,

$$(1 + n)^x = \sum_{k=0}^{x} \binom{x}{k} n^k = 1 + nx + \binom{x}{2} n^2 + \text{higher powers of } n$$

This indicates that

$$(1 + n)^x = 1 + nx \pmod{n^2}$$

Therefore, if

$$y = (1 + n)^x \mod n^2$$

then

$$x = \frac{y - 1}{n} \pmod{n}$$

Thus

$$L((1 + n)^x (mod\ n^2)) = x \pmod{n}$$

for any $x \in \mathbb{Z}_n$.

Therefore, when $g = n + 1$, we have

$$
\begin{aligned}
L(c^\lambda (mod\ n^2)) \cdot \mu &= L((g^m r^n)^\lambda (mod\ n^2)) \cdot \lambda^{-1} \\
&= L((g^{m\lambda} (mod\ n^2)) \cdot \lambda^{-1} \\
&= \lambda \cdot m \cdot \lambda^{-1} = m (mod\ n)
\end{aligned}
$$

Paillier Example: An example of the Paillier encryption scheme with small parameters is shown as follows.

For ease of calculations, the example will choose small primes, to create a small n. Let

$$p = 7, q = 11$$

then

$$n = pq = 7 \cdot 11 = 77$$

Next, an integer g must be selected from $\mathbb{Z}_{n^2}^*$, such that the order of g is a multiple of n in $\mathbb{Z}_{n^2}$. If we randomly choose the integer

$$g = 5652$$

then all necessary properties, including the yet to be specified condition, are met, as the order of g is $2310 = 30 \cdot 77$ in $\mathbb{Z}_{n^2}$. Thus, the public key for the example will be

$$(n, g) = (77, 5652)$$

To encrypt a message

$$m = 42$$

where $m \in \mathbb{Z}_n$, choose a random

$$r = 23$$

where r is a nonzero integer and $r \in \mathbb{Z}_n$.

Compute

$$c = g^m r^n (mod \; n^2)$$
$$= 5652^{42} \cdot 23^{77} (mod \; 5929)$$
$$= 4624 (mod \; 5929)$$

To decrypt the ciphertext c, compute

$$\lambda = lcm(6, 10) = 30$$

Define $L(u) = (u - 1)/n$, compute

$$k = L(g^\lambda (mod \; n^2))$$
$$= L(5652^{30} (mod \; 5929))$$
$$= L(3928)$$
$$= (3928 - 1)/77$$
$$= 3927/77$$
$$= 51$$

Compute the inverse of k,

$$\mu = k^{-1} (mod \; n)$$
$$= 51^{-1} = 74 (mod \; 77)$$

Compute

$$m = L(c^\lambda mod \, n^2) \cdot \mu (mod \; n)$$
$$= L(4624^{30} (mod \; 5929)) \cdot 74 (mod \; 77)$$
$$= L(4852) \cdot 74 (mod \; 77)$$
$$= 42$$

Homomorphic Properties: A notable feature of the Paillier scheme is its homomorphic properties. Given two ciphertexts $E(m_1, pk) = g^{m_1} r_1^n (mod \ n^2)$ and $E(m_2, pk) = g^{m_2} r_2^n (mod \ n^2)$, where r_1 and r_2 are randomly chosen from $\mathbb{Z}_n^*$, we have

- Homomorphic Addition of Plaintexts
    The product of two ciphertexts will decrypt to the sum of their corresponding plaintexts, i.e.,

$$D(E(m_1, pk) \cdot E(m_2, pk) \ (mod \ n^2)) = m_1 + m_2 (mod \ n)$$

because

$$\begin{aligned}
E(m_1, pk) \cdot E(m_2, pk) &= (g^{m_1} r_1^n)(g^{m_2} r_2^n) \ (mod \ n^2) \\
&= g^{m_1 + m_2} (r_1 r_2)^n (mod \ n^2) \\
&= E(m_1 + m_2, pk)
\end{aligned}$$

The product of a ciphertext with a plaintext raising g will decrypt to the sum of the corresponding plaintexts, i.e.,

$$D(E(m_1, pk) \cdot g^{m_2} (mod \ n^2)) = m_1 + m_2 (mod \ n)$$

because

$$\begin{aligned}
E(m_1, pk) \cdot g^{m_2} &= (g^{m_1} r_1^n) g^{m_2} \ (mod \ n^2) \\
&= g^{m_1 + m_2} r_1^n (mod \ n^2) \\
&= E(m_1 + m_2, pk)
\end{aligned}$$

- Homomorphic Multiplication of Plaintexts
    An encrypted plaintext raised to the power of another plaintext will decrypt to the product of the two plaintexts, i.e.,

$$D(E(m_1, pk)^{m_2} (mod \ n^2)) = m_1 m_2 (mod \ n)$$

because

$$\begin{aligned}
E(m_1, pk)^{m_2} &= (g^{m_1} r_1^n)^{m_2} \ (mod \ n^2) \\
&= g^{m_1 m_2} (r_1^{m_2})^n (mod \ n^2) \\
&= E(m_1 m_2, pk)
\end{aligned}$$

More generally, an encrypted plaintext raised to a constant k will decrypt to the product of the plaintext and the constant, i.e.,

$$D(E(m_1, pk)^k (mod\ n^2)) = km_1 (mod\ n)$$

However, given the Paillier encryptions of two messages, there is no known way to compute an encryption of the product of these messages without knowing the private key.

Paillier Security: The Paillier encryption scheme provides semantic security against chosen-plaintext attacks (IND-CPA). The ability to successfully distinguish the challenge ciphertext essentially amounts to the ability to decide composite residuosity. The semantic security of the Paillier encryption scheme was proved under the decisional composite residuosity (DCR) assumption—the DCR problem is intractable.

The DCR problem states as follows: Given a composite N and an integer z, it is hard to decide whether z is a N-residue modulo N^2 or not, i.e., whether there exists y such that

$$z = y^n (mod\ n^2)$$

Because of the homomorphic properties, the Paillier encryption scheme, however, is malleable and therefore does not protect against adaptive chosen-ciphertext attacks (IND-CCA2). Usually in cryptography the notion of malleability is not seen as an "advantage," but under certain applications such as secure electronic voting and threshold cryptosystems, this property may indeed be necessary.

Paillier and Pointcheval [12] however went on to propose an improved cryptosystem that incorporates the combined hashing of message m with random r. Similar in intent to the Cramer–Shoup cryptosystem, the hashing prevents an attacker, given only c, from being able to change m in a meaningful way. Through this adaptation the improved scheme can be shown to be IND-CCA2 secure in the random oracle model.

2.5 Boneh–Goh–Nissim Encryption Scheme

Boneh–Goh–Nissim encryption scheme [2], BGN scheme by brevity, resembles the Paillier [11] and the Okamoto–Uchiyama [10] encryption schemes. The BGN scheme was the first to allow both additions and multiplications with a constant-size ciphertext. The multiplication is possible due to the fact that pairings can be defined for elliptic curves.

Let G_1, G_2 be additive groups and G_T a multiplicative group, all of prime order p. Let $P \in G_1, Q \in G_2$ be generators of G_1 and G_2, respectively.

A pairing is a map

$$e : G_1 \times G_2 \rightarrow G_T$$

for which the following holds:

1. Bilinearity: $\forall a, b \in \mathbb{Z}_p^*$:

$$e(P^a, Q^b) = e(P, Q)^{ab}$$

2. Non-degeneracy: $e(P, Q) \neq 1$.
3. For practical purposes, e has to be computable in an efficient manner.

In cases when $G_1 = G_2 = G$, the pairing is called symmetric. If, furthermore, G is cyclic, the map e will be commutative; that is, for any $P, Q \in G$, we have

$$e(P, Q) = e(Q, P)$$

This is because for a generator $g \in G$, there exist integers p, q such that $P = g^p$ and $Q = g^q$. Therefore

$$e(P, Q) = e(g^p, g^q) = e(g, g)^{pq} = e(g^q, g^p) = e(Q, P)$$

On the basis of pairing, BGN scheme can be described by three algorithms—key generation, encryption, and decryption algorithms—as follows:

Key Generation: Given a security parameter $\lambda \in \mathbb{Z}^+$, generate a tuple (q_1, q_2, G, G_1, e), where q_1 and q_2 are two distinct large primes, G is a cyclic group of order $q_1 q_2$, and e is a pairing map $e : G \times G \rightarrow G_1$. Let $N = q_1 q_2$. Pick up two random generators g, u from G and set $h = u^{q_2}$. Then h is a random generator of the subgroup of G of order q_1. The public key is $PK = \{N, G, G_1, e, g, h\}$. The private key $SK = q_1$.

Encryption: Assume the message space consists of integers in the set $\{0, 1, \cdots, T\}$ with $T < q_2$. We encrypt bits in which case $T = 1$. To encrypt a message m using the public key PK, pick a random r from $\{1, 2, \cdots, N\}$ and compute

$$C = g^m h^r \in G \tag{2.18}$$

Output C as the ciphertext.

Decryption: To decrypt a ciphertext C using the private key $SK = q_1$, observe that

$$C^{q_1} = (g^m h^r)^{q_1} = (g^{q_1})^m \tag{2.19}$$

To recover the message m, it suffices to compute the discrete logarithm of C^{q_1} to the base g^{q_1}. Since $0 \leq m \leq T$, this takes expected time $O(\sqrt{T})$ using Pollard's lambda method [9].

Homomorphic Properties: The BGN scheme is clearly additively homomorphic. Let $PK = \{N, G, G_1, e, g, h\}$ be a public key. Given two ciphertexts $C_1 = g^{m_1} h^{r_1} \in G, C_2 = g^{m_2} h^{r_2} \in G$ of messages $m_1, m_2 \in \{0, 1, \cdots, T\}$ respectively, anyone can create a uniformly distributed encryption of $m_1 + m_2 (mod\ N)$

by computing the product

$$C = C_1 C_2 h^r \tag{2.20}$$

for a random r in $\{1, 2, \cdots, N - 1\}$, because

$$C_1 C_2 h^r = (g^{m_1} h^{r_1})(g^{m_2} h^{r_2}) h^r = g^{m_1 + m_2} h^{r_1 + r_2 + r}$$

is an encryption of $m_1 + m_2$.

More importantly, anyone can multiply two encrypted messages once using the bilinear map. Let

$$g_1 = e(g, g)$$

and

$$h_1 = e(g, h)$$

then g_1 is of order N and h_1 is of order q_1. There is some (unknown) $\alpha \in \mathbb{Z}$ such that

$$h = g^{\alpha q_2}$$

Suppose that we are given two ciphertexts $C_1 = g^{m_1} h^{r_1} \in G$ and $C_2 = g^{m_2} h^{r_2} \in G$. To build an encryption of the product $m_1 m_2 (mod\ N)$, (1) pick a random $r \in \mathbb{Z}_N$, and (2) let

$$C = e(C_1, C_2) h_1^r \in G_1 \tag{2.21}$$

We have

$$
\begin{aligned}
C &= e(C_1, C_2) h_1^r \\
&= e(g^{m_1} h^{r_1}, g^{m_2} h^{r_2}) h_1^r \\
&= e(g^{m_1 + \alpha q_2 r_1}, g^{m_2 + \alpha q_2 r_2}) h_1^r \\
&= e(g, g)^{(m_1 + \alpha q_2 r_1)(m_2 + \alpha q_2 r_2)} h_1^r \\
&= e(g, g)^{m_1 m_2 + \alpha q_2 (m_1 r_2 + m_2 r_1 + \alpha q_2 r_1 r_2)} h_1^r \\
&= e(g, g)^{m_1 m_2} h_1^{r + m_1 r_2 + m_2 r_1 + \alpha q_2 r_1 r_2}
\end{aligned}
$$

where $r + m_1 r_2 + m_2 r_1 + \alpha q_2 r_1 r_2$ is distributed uniformly in $\mathbb{Z}_N$. Thus C is a uniformly distributed encryption of $m_1 m_2 (mod\ N)$, but in G_1 rather than G. We note that the BGN scheme is still additively homomorphic in G_1.

BGN Example: We will demonstrate the operation of the BGN scheme with a small example. First we choose two distinct prime numbers

$$q_1 = 7, q_2 = 11$$

and compute the product

$$N = q_1 q_2 = 77$$

Next we construct an elliptic curve group with order N that has an associated bilinear map e. The equation for the elliptic curve is

$$y^2 = x^3 + x$$

and is defined over the field F_q for some prime $q = 3 \bmod 4$. In this example, we set

$$q = 307$$

Therefore, the curve is supersingular with $\#(E(q)) = q + 1 = 308$ rational points, which contains a subgroup G with the order $N = 77$ (=308/4).

Within the group G, we choose two random generators

$$g = [182, 240], u = [28, 262]$$

where these two generators have order N, and compute

$$h = u^{q_2} = [28, 262]^{11} = [99, 120]$$

where h has order $q_1 = 7$.

We compute the ciphertext of a message

$$m = 2$$

Take $r = 5$ and compute

$$C = g^m h^r = [182, 240]^2 \oplus [99, 120]^5 = [256, 265]$$

To decrypt we first compute

$$\hat{g} = g^{q_1} = [182, 240]^7 = [146, 60]$$

and

$$C^{q_1} = [256, 265]^7 = [299, 44]$$

Now we find the discrete logarithm by iterating through all the powers of $\hat{g} = g^{q_1}$ as follows:

$$\hat{g}^1 = [146, 60]$$
$$\hat{g}^2 = [299, 44]$$
$$\hat{g}^3 = [272, 206]$$
$$\hat{g}^4 = [191, 151]$$
$$\hat{g}^5 = [79, 171]$$
$$\hat{g}^6 = [79, 136]$$
$$\hat{g}^7 = [191, 156]$$
$$\hat{g}^8 = [272, 101]$$
$$\hat{g}^9 = [299, 263]$$
$$\hat{g}^{10} = [146, 247]$$
$$\hat{g}^{11} = \infty$$

Observe that $\hat{g}^2 = C^{q_1}$. Therefore, decryption of the ciphertext equals 2, which is the same as the original message.

BGN Security: The BGN encryption scheme has been proved to be semantically secure on basis of the subgroup decision problem in [2]. The subgroup decision (SD) problem is stated as follows.

Given a group G of composite order $n = pq$, where p, q are distinct (unknown) primes, and generators $g_p \in G_p$ and $g \in G$, distinguish between whether an element x is a random element of the subgroup G_p or a random element of the full group G.

Gjosteen [6] has undertaken an extensive survey of such problems, which he calls subgroup membership problems. For example, the quadratic residuosity problem is a subgroup membership problem: if we let $N = pq$ be a product of two distinct primes and define the group G to be the group of elements of $\mathbb{Z}_N^*$ with Jacobi symbol 1, the problem is to determine whether a given element in G lies in the subgroup of squares in G.

Boneh, Goh, and Nissim [2] defined their SD problem for pairs of groups (G, G_1) of composite order $N = pq$ for which there exists a nondegenerate bilinear map, or pairing, $e : G \times G \rightarrow G_1$. The problem is to determine whether a given element $x \in G$ is in the subgroup of order p. Note that if g generates G, then $e(g, x)$ is a challenge element for the same problem in G_1; thus if the SD problem is infeasible in G, then it is in G_1 as well.

Freeman [5] developed an abstract framework that encompasses the key properties of bilinear groups of composite order that are required to construct secure pairing-based cryptosystems and showed how to use prime-order elliptic curve groups to construct bilinear groups with the same properties. In particular, he defined a generalized version of the subgroup decision problem and give explicit constructions of bilinear groups in which the generalized subgroup decision assumption follows from the decision Diffie–Hellman assumption, the decision linear assumption, and/or related assumptions in prime-order groups.

References

1. M. Abdalla, M. Bellare, P. Rogaway, DHAES: an encryption scheme based on the Diffie–Hellman problem. Submission to IEEE P1363a, 1998. http://www.di.ens.fr/~mabdalla/papers/dhes.pdf
2. D. Boneh, E. Goh, K. Nissim, Evaluating 2-DNF formulas on ciphertexts, in *Proceedings of Theory of Cryptography, TCC'05*, 2005, pp. 325–341
3. R. Cramer, V. Shoup, A practical public key cryptosystem provably secure against adaptive chosen ciphertext attack, in *Proceedings of Advances in Cryptology, CRYPTO'98*, 1998, pp. 13–25
4. T. ElGamal, A public-key cryptosystem and a signature scheme based on discrete logarithms. IEEE Trans. Inf. Theory **31**(4), 469–472 (1985)
5. D.M. Freeman, Converting pairing-based cryptosystems from composite-order groups to prime-order groups, in *Proceedings of Advances in Cryptology, EUROCRYPT'10*, 2010, pp. 44–61
6. K. Gjosteen, Subgroup membership problems and public key cryptosystems, Dissertation, Norwegian University of Science and Technology, 2004
7. S. Goldwasser, S. Micali, Probabilistic encryption and how to play mental poker keeping secret all partial information, in *Proceedings of 14th Symposium on Theory of Computing*, 1982, pp. 365–377
8. S. Goldwasser, S. Micali, Probabilistic encryption. J. Comput. Syst. Sci. **28**(2), 270–299 (1984)
9. A. Menezes, P. van Oorschot, S. Vanstone, *Handbook of Applied Cryptography*. CRC Press, 1996
10. T. Okamoto, S. Uchiyama, A new public-key cryptosystem as secure as factoring, in *Proceedings of Advances in Cryptology, EUROCRYPT'98*, 1998, pp. 308–318
11. P. Paillier, Public key cryptosystems based on composite degree residue classes, *Proceedings of Advances in Cryptology, EUROCRYPT'99*, 1999, pp. 223–238
12. P. Paillier, D. Pointcheval, Efficient public-key cryptosystems provably secure against active adversaries, in *Proceedings of Advances in Cryptology, ASIACRYPT'99*, 1999, pp. 165–179

Chapter 3
Fully Homomorphic Encryption

Abstract Homomorphic encryption is a very useful tool with a number of attractive applications. However, the applications are limited by the fact that only one operation is possible (usually addition or multiplication in the plaintext space) to be able to manipulate the plaintext by using only the ciphertext. What would really be useful is to be able to utilize both addition and multiplication simultaneously. This would permit more manipulation of the plaintext by modifying the ciphertext. In fact, this would allow one without the secret key to compute any efficiently computable function on the plaintext when given only the ciphertext. In this chapter, we introduce fully homomorphic encryption (FHE) techniques, which allow one to evaluate both addition and multiplication of plaintext, while remaining encrypted. The concept of FHE was introduced by Rivest [14] under the name privacy homomorphisms. The problem of constructing a scheme with these properties remained unsolved until 2009, when Gentry [6] presented his breakthrough result. His scheme allows arbitrary computation on the ciphertexts and it yields the correct result when decrypted. This chapter begins with an introduction of FHE model and definitions, followed by the construction of FHE scheme over integers.

3.1 Fully Homomorphic Encryption Definition

Fully homomorphic encryption can be considered as ring homomorphism. In mathematics, a ring is a set R equipped with two operations $+$ and $\times$ satisfying the following eight axioms, called the ring axioms.

R is an abelian group under addition, meaning:

1. $(a + b) + c = a + (b + c)$ for all a, b, c in R ($+$ is associative).
2. There is an element 0 in R such that $a + 0 = a$ and $0 + a = a$ (0 is the additive identity).
3. For each a in R there exists $-a$ in R such that $a + (-a) = (-a) + a = 0$ ($-a$ is the additive inverse of a).
4. $a + b = b + a$ for all a and b in R ($+$ is commutative).

R is a monoid under multiplication, meaning:

© Xun Yi, Russell Paulet, Elisa Bertino 2014
X. Yi et al., *Homomorphic Encryption and Applications*, SpringerBriefs
in Computer Science, DOI 10.1007/978-3-319-12229-8__3

5. $(a \cdot b) \cdot c = a \cdot (b \cdot c)$ for all a, b, c in R ($\cdot$ is associative).
6. There is an element 1 in R such that $a \cdot 1 = a$ and $1 \cdot a = a$ (1 is the multiplicative identity).

Multiplication distributes over addition:

7. $a \cdot (b + c) = (a \cdot b) + (a \cdot c)$ for all a, b, c in R (left distributivity).
8. $(b + c) \cdot a = (b \cdot a) + (c \cdot a)$ for all a, b, c in R (right distributivity).

A ring homomorphism is a function between two rings which respects the structure. More explicitly, if R and S are two rings, then a ring homomorphism is a function

$$f : R \to S$$

such that

$$f(a + b) = f(a) + f(b) \tag{3.1}$$
$$f(a \cdot b) = f(a) \cdot f(b) \tag{3.2}$$

for all a and b in R.

Let us see an example of ring homomorphism. Consider the function

$$f : Z_2 \to Z_2$$

given by

$$f(x) = x^2$$

where $x = 0$ or 1.

First,

$$f(x + y) = (x + y)^2 = x^2 + 2xy + y^2 = x^2 + y^2 = f(x) + f(y)$$

where $2xy = 0$ because 2 times anything is 0 in Z_2.

Next,

$$f(xy) = (xy)^2 = x^2 y^2 = f(x) f(y)$$

The second equality follows from the fact that Z_2 is commutative. Thus, f is a ring homomorphism.

Let (P, C, K, E, D) be a encryption scheme, where P, C are the plaintext and ciphertext spaces, K is the key space, and E, D are the encryption and decryption algorithms. Assume that the plaintexts form a ring $(P, \oplus_p, \otimes_p)$ and the ciphertexts form a ring $(C, \oplus_c, \otimes_c)$; then the encryption algorithm E is a map from the ring P to the ring C, i.e., $E_k : P \to C$, where $k \in K$ is either a secret key (in the secret key cryptosystem) or a public key (in the public-key cryptosystem).

For all a and b in P and k in K, if

$$E_k(a) \oplus_c E_k(b) = E_k(a \oplus_p b) \tag{3.3}$$

$$E_k(a) \otimes_c E_k(b) = E_k(a \otimes_p b) \tag{3.4}$$

the encryption scheme is **fully homomorphic**.

3.2 Overview of Fully Homomorphic Encryption Schemes

Craig Gentry [6, 7], using lattice-based cryptography, showed the first fully homomorphic encryption scheme as announced by IBM on 25 June 2009. His scheme supports evaluations of arbitrary depth circuits. His construction starts from a somewhat homomorphic encryption scheme using ideal lattices that is limited to evaluating low-degree polynomials over encrypted data. It is limited because each ciphertext is noisy in some sense, and this noise grows as one adds and multiplies ciphertexts, until ultimately the noise makes the resulting ciphertext indecipherable. He then shows how to modify this scheme to make it bootstrappable—in particular, he shows that by modifying the somewhat homomorphic scheme slightly, it can actually evaluate its own decryption circuit, a self-referential property. Finally, he shows that any bootstrappable somewhat homomorphic encryption scheme can be converted into a fully homomorphic encryption through a recursive self-embedding.

In the particular case of Gentry's ideal-lattice-based somewhat homomorphic scheme, this bootstrapping procedure effectively "refreshes" the ciphertext by reducing its associated noise so that it can be used thereafter in more additions and multiplications without resulting in an indecipherable ciphertext. Gentry based the security of his scheme on the assumed hardness of two problems: certain worst-case problems over ideal lattices and the sparse (or low-weight) subset sum problem.

Regarding performance, ciphertexts in Gentry's scheme remain compact insofar as their lengths do not depend at all on the complexity of the function that is evaluated over the encrypted data. The computational time only depends linearly on the number of operations performed. However, the scheme is impractical for many applications, because ciphertext size and computation time increase sharply as one increases the security level. To obtain 2^k security against known attacks, the computation time and ciphertext size are high-degree polynomials in k. Stehle and Steinfeld [16] reduced the dependence on k substantially. They presented optimizations that permit the computation to be only quasi-$k^{3.5}$ per Boolean gate of the function being evaluated.

In 2009, Marten van Dijk, Craig Gentry, Shai Halevi, and Vinod Vaikuntanathan [5] presented a second fully homomorphic encryption scheme, which uses many of the tools of Gentry's construction, but which does not require ideal lattices. Instead, they show that the somewhat homomorphic component of Gentry's ideal lattice-based scheme can be replaced with a very simple somewhat homomorphic scheme that uses integers. The scheme is therefore conceptually simpler than Gentry's ideal lattice scheme, but has similar properties with regard to homomorphic operations and efficiency.

In 2010, Nigel P. Smart and Frederik Vercauteren [15] presented a fully homomorphic encryption scheme with smaller key and ciphertext sizes. The Smart–Vercauteren scheme follows the fully homomorphic construction based on ideal lattices given by Gentry [6]. It also produces a fully homomorphic scheme from a somewhat homomorphic scheme. For somewhat homomorphic scheme, the public and the private keys consist of two large integers (one of which shared by both the public and the private keys), and the ciphertext consists of one large integer. The Smart–Vercauteren scheme has smaller ciphertext and reduced key size than Gentry's scheme based on ideal lattices. Moreover, the scheme also allows efficient fully homomorphic encryption over any field of characteristic two. However, the major problem with this scheme is that the key generation method is very slow. This scheme is still not fully practical.

At the rump session of Eurocrypt 2011, Craig Gentry and Shai Halevi [8] presented a working implementation of fully homomorphic encryption (i.e., the entire bootstrapping procedure) together with performance numbers.

Recently, Coron, Naccache, and Tibouchi [4] proposed a technique allowing to reduce the public-key size of the van Dijk et al. scheme to 600 KB. In April 2013 the HElib [9] was released, via GitHub, to the open source community which implements the Brakerski-Gentry-Vaikuntanathan (BGV) homomorphic encryption scheme [1], along with many optimizations to make homomorphic evaluation runs faster.

3.3 Somewhat Homomorphic Encryption Scheme over Integers

Although interesting from a theoretical standpoint, the lattice-based construction is difficult to describe. We now move to a scheme that is easier to understand. It can be seen as the integer-based version of the lattice version. That is, we can embed an ideal into an integer ring, and if the parameters are set correctly, the scheme can be considered secure (against known attacks). As a bonus, the bootstrapping procedure is easier to understand and describe in greater detail, since it requires zero background with lattices.

3.3.1 Secret Key Somewhat Homomorphic Encryption

We begin with the description of the secret key integer-based somewhat homomorphic encryption scheme [5]. The scheme is surprisingly simple, and we can construct very complex functionality from it.

Key Generation KeyGen: The secret key is an odd integer, chosen from some interval $p \in [2^{\eta-1}, 2^{\eta}]$.

Encryption Encrypt(pk, m): To encrypt a bit $m \in \{0, 1\}$, set the ciphertext as an integer whose residue mod p has the same parity as the plaintext. Namely, set

$$c = pq + 2r + m \tag{3.5}$$

where the integers q, r are chosen at random in some other prescribed intervals, such as $2r$ is smaller than $p/2$ in absolute value.

Decryption Decrypt(p, c): Given a ciphertext c and the secret key p, output

$$m = (c(mod\ p))(mod\ 2) \tag{3.6}$$

The decryption equation holds because

$$(c(mod\ p))(mod\ 2) = (pq + 2r + m(mod\ p))(mod\ 2)$$
$$= 2r + m(mod\ 2)$$
$$= m$$

For example, suppose that $p = 17$; let us encrypt $m = 1$ as follows:

$$c = pq + 2r + m = 17 \cdot 2 + 2 \cdot 0 + 1 = 39$$

where $q = 2, r = 0$.
It is easy to see that

$$(c(mod\ p))(mod\ 2) = (39(mod\ 17))(mod\ 2)$$
$$= 1(mod\ 2) = 1$$

Fully Homomorphic Property: Given two ciphertext $c_1 = pq_1 + 2r_1 + m_1$ and $c_2 = pq_2 + 2r_2 + m_2$, we have

$$c_1 + c_2 = (q_1 + q_2)p + 2(r_1 + r_2) + (m_1 + m_2) \tag{3.7}$$
$$c_1 \cdot c_2 = (pq_1q_2 + 2q_1r_2 + 2q_2r_1 + m_1q_2 + m_2q_1)p$$
$$+ 2(2r_1r_2 + m_1r_2 + m_2r_1) + m_1m_2 \tag{3.8}$$

When

$$r_1 + r_2 < p/2$$
$$2r_1r_2 + m_1r_2 + m_2r_1 < p/2$$

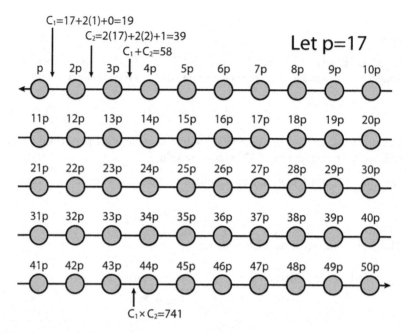

Fig. 3.1 An example of homomorphic addition and multiplication

we have

$$(c_1 + c_2(mod\ p))(mod\ 2) = m_1 + m_2$$
$$(c_1 \cdot c_2(mod\ p))(mod\ 2) = m_1 m_2$$

Therefore, this scheme has the fully homomorphic property.

For example, set $p = 17$, $m_1 = 0$, and $m_2 = 1$. Then compute ciphertexts as

$$c_1 = p \cdot 1 + 2 \cdot 1 + 0 = 19$$
$$c_2 = p \cdot 2 + 2 \cdot 2 + 1 = 39$$

where $q_1 = 1, r_1 = 1, q_2 = 2, r_2 = 2$. Figure 3.1 plots these points on the number line. In addition to the position of the ciphertexts, it also shows the sum and product.

It is easy to verify that

$$(c_1 + c_2(mod\ p))(mod\ 2) = (58(mod\ 17))(mod\ 2)$$
$$= 7(mod\ 2)$$
$$= 1 = 0 + 1 = m_1 + m_2$$
$$(c_1 \cdot c_2(mod\ p))(mod\ 2) = (741(mod\ 17))(mod\ 2)$$
$$= 10(mod\ 2)$$
$$= 0 = 0 \cdot 1 = m_1 \cdot m_2$$

However, when we use the fully homomorphic property to evaluate a Boolean function $f(x_1, x_2, \cdots, x_n)$ where $x_i \in \{0, 1\}$, given c_i, the encryption of x_i, for $i = 1, 2, \cdots, n$, it is noticed in Eqs. (3.7) and (3.8) that

$$r_1 + r_2 \geq max(r_1, r_2)$$

$$2r_1r_2 + m_1r_2 + m_2r_1 \geq max(r_1, r_2)$$

that is, the size of the noise component r in the resulted ciphertext is increasing with the number of the additions and multiplications in the Boolean function. Once

$$r_1 + r_2 > p/2$$

$$2r_1r_2 + m_1r_2 + m_2r_1 > p/2$$

the decryption of $f(c_1, c_2, \cdots, c_n)$ may not be $f(x_1, x_2, \cdots, x_n)$. Therefore, this scheme can be only used to evaluate low-degree Boolean functions over encrypted data. This is why this scheme is called somewhat homomorphic encryption scheme.

If we choose $r \approx 2^n$, $p \approx 2^{n^2}$, and $q \approx 2^{n^5}$, the somewhat encryption scheme can compute polynomials of degree $\approx n$ before the noise grows too large.

Security: The security of this scheme can be reduced to the hardness of the approximate integer greatest common divisor (approximate GCD) problem [10]. As an example, we explain this in the more specific and familiar case of greatest common divisors. If we are given two integers a and b we can clearly find their GCD, d say, in polynomial time. If d is in some sense large then it may be possible to incur some additive error on either of the inputs a and b, or both, and still recover this GCD. This is what we refer to as an approximate common divisor problem. Of course if there is too much error incurred on the inputs, the algorithm may well not be able to discern the GCD d we had initially over some other approximate divisors d (e.g., they may all leave residues of similar magnitude when dividing a and b). In this sense, the problem is similar to those found in error correcting codes.

Continuing this error correcting code analogy we can state the problem from the standpoint of the design of the decoding algorithm, i.e., we wish to create an algorithm which is given two inputs a_0 and b_0 and bounds X, Y, and M for which one is assured that $d|(a_0 + x_0)$ and $d|(b_0 + y_0)$ for some $d > M$ and x_0, y_0 satisfying $|x_0| \leq X, |y_0| \leq Y$. The output of the algorithm should be the common divisor d, or all of the possible ones if more than one exists.

Howgrave-Graham analyzed the (approximate GCD) problem in [10]. The problem is believed to be a hard problem in lattice theory.

With a judicious choice of parameters (e.g., $r \approx 2^{\sqrt{\eta}}$ and $q \approx 2^{\eta^3}$), the secret key somewhat homomorphic encryption scheme is even secure.

3.3.2 Public-Key Somewhat Homomorphic Encryption

The secret key somewhat homomorphic encryption needs the secret key p to encrypt a message. Now we describe a public-key somewhat homomorphic encryption scheme [5] that allows encryption without the knowledge of the secret p.

Parameters: The scheme has many parameters, controlling the number of integers in the public key and the bit-length of the various integers. Specifically, we use the following four parameters (all polynomial in the security parameter λ):

γ is the bit-length of the integers in the public key;
η is the bit-length of the secret key (which is the hidden approximate GCD of all the public-key integers);
ρ is the bit-length of the noise (i.e., the distance between the public-key elements and the nearest multiples of the secret key);
τ is the number of integers in the public key.

These parameters must be set under some constraints [5]. A convenient parameter set to keep in mind is $\rho = \lambda, \rho' = 2\lambda, \eta = O(\lambda^2), \gamma = O(\lambda^5)$, and $\tau = \gamma + \lambda$. The setting results in a scheme with complexity $O(\lambda^{10})$.

Key Generation   KeyGen(λ): Choose a random η-bit odd integer p as the private key. Using the private key, generate the public key as

$$x_i = pq_i + r_i \tag{3.9}$$

where $q_i \in \mathbb{Z} \bigcap [0, 2^\gamma / p)$ and $r_i \in \mathbb{Z} \bigcap (-2^\rho, 2^\rho)$ are chosen randomly, for $i = 0, 1, \cdots, \tau$. Relabel so that x_0 is the largest. Restart unless x_0 is odd and $x_0 (mod \ p)$ is even. The public key is

$$pk = < x_0, x_2, \cdots, x_\tau >$$

Encryption   Encrypt(pk, m): Given $m \in \{0, 1\}$ and the public key pk, choose a random subset $S \subseteq \{1, 2, \cdots, \tau\}$ and a random integer $r \in (2^{-\rho'}, 2^{\rho'})$, and output

$$c = (m + 2r + 2\sum_{i \in S} x_i)(mod \ x_0) \tag{3.10}$$

Decryption   Decrypt(sk, c): Given the ciphertext c and the private key p, output

$$m = (c(mod \ p))(mod \ 2) \tag{3.11}$$

Recall that

$$c(mod \ p) = c - p \cdot [c/p]$$

where $[a]$ denotes the rounding to the nearest integer. As p is odd, we can instead decrypt using the formula

$$m = (c - p \cdot [c/p])(mod\ 2)$$
$$= (c(mod\ 2)) \oplus ([c/p](mod\ 2)) \tag{3.12}$$

where $p(mod\ 2) = 1$.

Example. Let the secret key be $p = 10001$. Based on p, we construct the public key as follows.

Set

$$[q_0, q_1, q_2, q_3] = [36, 27, 34, 6]$$
$$[r_0, r_1, r_2, r_3] = [8, 5, 4, 2]$$

We then compute the public key pk as the vector

$$[x_0, x_1, x_2, x_3] = [360044, 270032, 340038, 60008]$$

where $x_i = q_i p + r_i$ and x_0 is the largest.

We now encrypt two messages $m_1 = 0, m_2 = 1$ using a random subset of the public key. Suppose that the subset is $S = [1, 3]$. We select a random integer $r = 31$ and encrypt m_1 as:

$$c_1 = m_1 + 2 \cdot r + 2 \cdot \sum_{i \in S} x_i$$
$$= 0 + 2 \cdot 31 + 2 \cdot (270032 + 60008)$$
$$= 660142$$

For the sake of compactness, it is useful to reduce the ciphertext by x_0 as

$$c_1' = c_1(mod\ x_0) = 300098(mod\ 360044)$$

We encrypt m_2 using the same process. We set $r = 11$ and $S = [2, 3]$.

$$c_2 = m_2 + 2 \cdot r + 2 \cdot \sum_{i \in S} x_i$$
$$= 1 + 2 \cdot 11 + 2 \cdot (340038 + 60008)$$
$$= 800115$$

Again, for the sake of compactness, we reduce the ciphertext by x_0 as

$$c_2' = c_2(mod\ x_0) = 80027(mod\ 360044)$$

As expected, these ciphertexts are decrypted correctly as shown by the following equations:

$$c_1' = 300098 = 68(mod\ p) = 0(mod\ 2)$$
$$c_2' = 80027 = 19(mod\ p) = 1(mod\ 2)$$

In addition,

$$c_1' + c_2' = 300098 + 80027 = 87(mod\ p) = 1(mod\ 2) = 0 + 1$$
$$c_1' \cdot c_2' = 300098 \cdot 80027 = 1292(mod\ p) = 0(mod\ 2) = 0 \cdot 1$$

Correctness: van Dijk et al. [5] provided the proof of correctness for the public-key somewhat homomorphic encryption scheme by some lemmas as follows.

Lemma 3.1 ([5]). *Let (sk, pk) be output by $\mathsf{KeyGen}(\lambda)$. Let $c = \mathsf{Encrypt}(pk, m)$ for $m \in \{0, 1\}$. Then $c = a \cdot p + (2b + m)$ for some integers a and b with $|2b + m| \le \tau 2^{\rho+3}$.*

Proof ([5]). By definition, $c = m + 2r + \sum_{i \in S} x_i (mod\ x_0)$. Since $|x_0| \ge |x_i|$ for $i = 1, 2, \cdots, \tau$, we have that

$$c = \left(m + 2r + \sum_{i \in S} x_i \right) + k \cdot x_0$$

for some $|k| \le \tau$.

For every i, there exist integers q_i and r_i with $|r_i| \le 2^\rho$ such that $x_i = q_i \cdot p + 2r_i$. We have

$$c = p \cdot \left(kq_0 + \sum_{i \in S} q_i \right) + \left(m + 2r + k \cdot 2r_0 + \sum_{i \in S} 2r_i \right)$$

Regarding the rightmost term, its parity is the same as m, and its absolute value is at most $(4\tau + 3)2^\rho < \tau 2^{\rho+3}$. $\square$

For a mod-2 arithmetic circuit (composed of mod-2 Add and Mult gates), let us consider its generalization to the integers, i.e., the same circuits with the Add and Mult gates applied to integers rather than to bits. A permitted circuit [5] is defined as one where for any $\alpha \ge 1$ and any set of integer inputs all less than $2^{\alpha(\rho'+2)}$ in absolute value, it holds that the generalized circuit's output has absolute value at most $2^{\alpha(\eta-4)}$.

Lemma 3.2 ([5]). *Let (sk, pk) be output by $\mathsf{KeyGen}(\lambda)$. Let C be a permitted circuit with t inputs and one output. For $i \in \{1, 2, \cdots, t\}$ and $m_i \in \{0, 1\}$. Let $c_i = \mathsf{Encrypt}(pk, m_i)$ and $m = C(m_1, m_2, \cdots, m_t)$ and $c = C'(c_1, c_2, \cdots, c_t)$*

where C' is the generalized circuit corresponding to C. Then $c = a \cdot p + (2b + m)$
for some integers a and b with $|2b + m| < p/8$.

Proof ([5]). Generally, we have

$$C'(c_1, c_2, \cdots, c_t) \in C'(2b_1 + m_1, \cdots, 2b_t + m_t) + p\mathbb{Z}$$

So $C'(2b_1 + m_1, \cdots, 2b_t + m_t)(mod\ p)$ has the same parity as $m = C(m_1, m_2, \cdots, m_t)$. We also have that

$$C'(2b_1 + m_1, \cdots, 2b_t + m_t) \le 2^\eta/16 \le p/8$$

by the definition of permitted circuits, since $|2b_i + m_i| < \tau 2^{\rho+3}$ by Lemma 3.1. $\square$

Based on Lemmas 3.1 and 3.2, we can see that for any permitted circuit C and any encryptions of inputs to that circuit, the integer output by the evaluation is of the form

$$c = a \cdot p + (2b + m)$$

with

$$|2b + m| \le p/8$$

where m is the plaintext that c is supposed to encrypt. Accordingly, we have

$$(c(mod\ p))(mod\ 2) = (2b + m)(mod\ 2) = m$$

Therefore, the public-key somewhat homomorphic encryption scheme can correctly evaluate any permitted circuit.

The definition of the permitted circuit is rather indirect. In particular, this definition does not give a good picture of what a permitted circuit looks like. By the triangle inequality, a k-fan-in Add gate clearly increases the magnitude of the integers by at most a factor of k. However, a 2-fan-in Mult gate may square the magnitude of the integers—i.e., double their bit-lengths. So, clearly, the main bottleneck is the multiplicative depth of the circuit, or the degree of the multivariate polynomial computed by the circuit.

Lemma 3.3 ([5]). *Let C be a Boolean circuit with t inputs and C^* be the associated integer circuit (where Boolean gates are replaced with integer operations). Let $f(x_1, x_2, \cdots, x_t)$ be the multivariate polynomial computed by C^* and d be its degree. If $|f| \cdot (2^{\rho'+2})^d \le 2^{\eta-4}$ (where $|f|$ is the ℓ_1 normal of the coefficient vector of f), then C is a permitted circuit.*

In particular, the somewhat homomorphic encryption scheme can handle f as long as

$$d \leq \frac{\eta - 4 - \log |f|}{\rho' + 2}$$

Security: Like the secret key homomorphic encryption scheme, the security of the public-key somewhat homomorphic encryption scheme is also based on approximate-GCD problem.

Consider the approximate-GCD instance $\{x_0, x_1, \cdots, x_t\}$ where $x_i = pq_i + r_i$. Known attacks on the approximate-GCD problem for two numbers include brute-forcing the reminders, continued fractions, and Howgrave-Graham's approximate-GCD algorithm [10].

A simple brute-force attack is to try to guess r_1 and r_2 and verify the guess with a GCD computation. Specifically, for $r_1', r_2' \in (2^{-\rho}, 2^\rho)$, set

$$x_1' = x_1 - r_1', x_2' = x_2 - r_2', p' = GCD(x_1', x_2')$$

If p' has η bits, output p' as a possible solution. The solution p will definitely be found by this technique, and for the parameter choices, where ρ is much smaller than η, the solution is likely to be unique. The running time of the attack is approximately $2^{2\rho}$.

Attacks for arbitrarily large values of t include lattice-based algorithms for simultaneous Diophantine approximate [11], Nguyen and Stern's orthogonal lattice [13], and extensions of Coppersmith's method to multivariate polynomials [2].

3.4   Fully Homomorphic Encryption Scheme over Integers

In this section, we describe the construction of a fully homomorphic encryption scheme given by van Dijk [5]. It is built on the somewhat homomorphic encryption scheme described in the last section and squashing the decryption circuit.

3.4.1   Squashed Encryption

Let κ, θ, Θ be three more parameters, which are a function of λ. We set $\kappa = \gamma\eta/\rho', \theta = \lambda$, and $\Theta = \omega(\kappa \cdot \log \lambda)$. For a secret key $sk^* = p$ and public key pk^* from the original somewhat homomorphic scheme, we add to the public key a set $y = \{y_1, y_2, \cdots, y_\Theta\}$ of rational numbers in [0,2) with κ bits of precision, such that there is a sparse subset $S \subset \{1, 2, \cdots, \Theta\}$ of size θ with

$$\sum_{i \in S} y_i \approx 1/p (mod\ 2)$$

Now the secret key is replaced by the indicator vector of the subset S. The encryption scheme is modified by van Dijk [5] as follows.

Key Generation   KeyGen(λ): Generate $sk^* = p$ and pk^* as before. Set $x_p = [2^\kappa/p]$, choose at random a Θ-bit vector $(s_1, s_2, \cdots, s_\Theta)$ with Hamming weight θ, and let $S = \{i : s_i = 1\}$.
Choose at random integer $u_i \in \mathbb{Z} \cap [0, 2^{\kappa+1})$, $i = 1, 2, \cdots, \Theta$, subject to the condition that

$$\sum_{i \in S} u_i = x_p (mod\ 2^{\kappa+1})$$

Set $y_i = u_i/2^\kappa$ and $y = \{y_1, y_2, \cdots, y_\Theta\}$. Hence, each y_i is a positive number smaller than 2, with κ bits of precision after the binary point. Also we have

$$\sum_{i \in S} y_i (mod\ 2) = (1/p) - \Delta_p$$

for some $|\Delta_p| < 2^{-\kappa}$ because

$$
\begin{aligned}
\sum_{i \in S} y_i &= \sum_{i \in S} u_i/2^\kappa \\
&= (x_p + \alpha \cdot 2^{\kappa+1})/2^\kappa \\
&= x_p/2^\kappa + \alpha \cdot 2 \\
&= [2^\kappa/p]/2^\kappa + \alpha \cdot 2 \\
&= (1/p - \Delta/2^\kappa) + \alpha \cdot 2 \\
&= 1/p - \Delta_p (mod\ 2)
\end{aligned}
$$

where $|\Delta| < 1$.
Output the secret key $sk = S$ and the public key $\{pk, y\}$.

Encryption   Encrypt(pk, c^*): Given a ciphertext c^*, for $i \in \{1, 2, \cdots, \Theta\}$, set

$$z_i = c^* \cdot y_i (mod\ 2)$$

keeping only $n = \lceil \log \theta \rceil + 3$ bits of precision after the binary point for each z_i.
Output both c^* and $z = \{z_1, z_2, \cdots, z_\Theta\}$.

Decryption   Decrypt(sk, c^*, z): Given the ciphertext c^*, z and the private key p, output

$$m = \left(c^* - \left[\sum_{i \in S} z_i \right] \right) (mod\ 2) \tag{3.13}$$

Correctness: van Dijk et al. [5] provided the proof of correctness for the squashed encryption by the following lemma.

Lemma 3.4 ([5]). *The squashed encryption scheme is correct for permitted polynomials. Moreover, for every ciphertext (c^*, z) that is generated by evaluating a permitted polynomial, it holds that $\sum_i s_i z_i$ is within 1/4 of an integer.*

Proof ([5]). Fix public and secret keys, generated with respect to security parameter λ with $y = \{y_1, y_2, \cdots, y_\Theta\}$ the rational numbers in the public key and $S = \{i : s_i = 1\}$ the secret key bits. Recall that the y_i were chosen so that $\sum_{i \in S} y_i (mod\ 2) = (1/p) - \Delta_p$ where $|\Delta_p| \le 2^{-\kappa}$. □

Fix a permitted polynomial $P(x_1, x_2, \cdots, x_t)$, given t ciphertexts $c_1, c_2, \cdots, c_t$, let $c^* = P(c_1, c_2, \cdots, c_t)$, we need to establish that

$$[c^*/p] = \left[\sum_i s_i z_i \right] (mod\ 2)$$

where z_i is computed as $c^* \cdot y_i (mod\ 2)$ with only $[\log_\theta] + 3$ bits of precision after the binary point, so $c^* \cdot y_i (mod\ 2) = z_i - \Delta_i$ with $|\Delta_i| \le 1/16\theta$. We have

$$(c^*/p) - \sum_i s_i z_i (mod\ 2) = (c^*/p) - \sum_i s_i (c^* \cdot y_i) + \sum_i s_i \Delta_i (mod\ 2)$$

$$= (c^*/p) - c^* \sum_i s_i y_i + \sum_i s_i \Delta_i (mod\ 2)$$

$$= (c^*/p) - c^* (1/p - \Delta_p) + \sum_i s_i \Delta_i (mod\ 2)$$

$$= c^* \cdot \Delta_p + \sum_i s_i \Delta_i (mod\ 2)$$

To establish the claim, observe that $|\sum_i s_i \Delta_i| \le \theta \cdot \frac{1}{16\theta} = 1/16$. Regarding $c^* \cdot \Delta_p$, recall that the output ciphertext c^* is obtained by evaluating the polynomial P on the input ciphertexts c_i (as if P was an integer polynomial). By the definition of a permuted polynomial, for any $\alpha > 1$, if P's inputs have magnitude at most $2^{\alpha(\rho'+2)}$, its output has magnitude at most $2^{\alpha(\eta-4)}$. In particular, when P's inputs are "fresh" ciphertext, which have magnitude at most 2^γ, P's output ciphertext c^* has magnitude at most $2^{\gamma(\eta-4)/(\rho'+2)} < 2^{\kappa-4}$. Thus, $|c^* \cdot \Delta_p| < 1/16$. Together, we have $c^* \cdot \Delta_p + \sum_i s_i \Delta_i (mod\ 2)$ that has magnitude at most 1/8 and therefore $[c^*/p] = [\sum_i s_i z_i](mod\ 2)$.

By definition, since c^* is a valid ciphertext output by a permitted polynomial, the value c^*/p is within $1/8$ of an integer. Together, it holds that $\sum_i s_i z_i$ is within $1/4$ of an integer.

Security: Like the original somewhat encryption scheme, the security of the squashed encryption scheme is still based on the approximate-GCD problem. Besides it, putting the hint $y = \{y_1, y_2, \cdots, y_\Theta\}$ in the public key induces another computational assumption, related to the sparse subset sum problem (SSSP) used by Gentry [6], and studied previously (sometimes under the name "low-weight" knapsack) in the context of server-aided cryptography [12] and in connection to Chor–Rivest cryptosystem [13].

The subset sum problem is an important problem in complexity theory and cryptography. The problem is this: given a set of integers, is there a nonempty subset whose sum is zero? For example, given the set $\{-7, -3, -2, 5, 8\}$, the answer is yes because the subset $\{-3, -2, 5\}$ sums to zero. The problem is NP-complete. An equivalent problem is this: given a set of integers and an integer s, does any nonempty subset sum to s? Subset sum can also be thought of as a special case of the knapsack problem.

Known attacks on the problem can be easily avoided by choosing θ large enough to avoid brute-force attacks (and improvements using time-space trade-off) and choosing Θ to be larger than $\omega(\log \lambda)$ times the bit-length of the rational numbers in the public key (which have length κ).

Example. Let the secret key be $p = 10001$. Set $\kappa = 24$ and

$$x_p = [2^{24}/p] = 1678$$

and choose at random 9-bit vector with Hamming weight 3, $s = \{0, 0, 1, 0, 0, 1, 0, 1, 0\}$, and let $S = \{3, 6, 8\}$. Choose at random integers $u_i \in \mathbb{Z} \cap [0, 2^{25})$, $i = 1, 2, \cdots, 9$ as follows:

$$u_1 = 281782$$

$$u_2 = 1892147$$

$$u_3 = 589103$$

$$u_4 = 487403$$

$$u_5 = 491831$$

$$u_6 = 1093482$$

$$u_7 = 293813$$

$$u_8 = 31873525$$

$$u_9 = 5718711$$

where

$$\sum_{i \in S} u_i = u_3 + u_6 + u_8$$

$$= 589103 + 1093482 + 31873525$$

$$= 1678 = x_p (mod\ 2^{25})$$

Set $y_i = u_i / 2^\kappa, i = 1, 2, \cdots, 9$ as follows:

$$y_1 = 0.0167955$$

$$y_2 = 0.1127807$$

$$y_3 = 0.0351133$$

$$y_4 = 0.0290515$$

$$y_5 = 0.0293154$$

$$y_6 = 0.0651766$$

$$y_7 = 0.0175126$$

$$y_8 = 1.8998101$$

$$y_9 = 0.3408617$$

where

$$\sum_{i \in S} y_i = y_3 + y_6 + y_8$$

$$= 0.0351133 + 0.0651766 + 1.8998101$$

$$= 2.0001 \approx 1/p (mod\ 2)$$

Given a ciphertext $c^* = 300098$ which is the encryption of 0, to re-encrypt c^*, we compute $z_i = c^* \cdot y_i (mod\ 2), i = 1, 2, \cdots, 9$ as follows:

$$z_1 = 0.295959$$

$$z_2 = 1.2625086$$

$$z_3 = 1.4311034$$

$$z_4 = 0.297047$$

$$z_5 = 1.4929092$$

$$z_6 = 1.3673068$$

$$z_7 = 1.4962348$$

$$z_8 = 1.2113898$$

$$z_9 = 1.9144466$$

Let $z = \{z_1, z_2, \cdots, z_9\}$; the re-encryption takes the form of (c^*, z).
For decryption, we compute

$$c^* - \left\lceil \sum_{i \in S} z_i \right\rfloor = 30098 - [z_3 + z_6 + z_8]$$

$$= 300098 - [1.4311034 + 1.3673068 + 1.2113898]$$

$$= 300098 - [4.0098]$$

$$= 300094 = 0 (mod\ 2)$$

The decryption result is the same as the original plaintext bit 0.

3.4.2 Bootstrappable Encryption

Now let us construct homomorphic encryption for circuits of any depth from somewhat homomorphic encryption, which is capable of evaluating just a little more than its own decryption circuit.

Definition 3.5 (Augmented Decryption Circuit [5]). Let ϵ be an encryption scheme, where decryption is implemented by a circuit that depends only on the security parameter.

For a given value of the security parameter λ, the set of augmented decryption circuits consists of two circuits; both take as input a secret key and two ciphertexts:

- The circuit decrypts both ciphertext and adds the resulting plaintext bits mod 2;
- The circuit decrypts both ciphertext and multiplies the resulting plaintext bits mod 2.

Definition 3.6 (Bootstrappable Encryption [5]). Let ϵ be a homomorphic encryption scheme. We say that ϵ is bootstrappable if its augmented decryption circuits are permitted circuits for every value of the security parameter λ.

Theorem 3.7 ([5]). *The squashed encryption scheme ϵ is bootstrappable.*

The details of the theorem proof can be found in [5].

To reduce the ciphertext size during evaluation, van Dijk et al. [5] added to the public key more elements of the form $x'_i = q'_i p + 2r_i$ where r_i is chosen as usual from the interval $(2^{-\rho}, 2^{\rho})$ but q_i are chosen much larger than for the other public-key elements. Specifically, for $i = 0, 1, \cdots, \gamma$, set

$$q_i' \in \mathbb{Z} \cap [2^{\gamma+i-1}/p, 2^{\gamma+i}/p), r_i \in \mathbb{Z} \cap (2^{-\rho}, 2^{\rho}), x_i' = 2(q_i' \cdot p + r_i)$$

thus getting $x_i' \in [2^{\gamma+i}, 2^{\gamma+i+1}]$.

During evaluation, every time we have a ciphertext that grows beyond 2^{γ}, we reduce its first modulo x_γ', then modulo $x_{\gamma-1}'$ and so on all the way down to x_0', at which point we again have a ciphertext of bit-length no more than γ.

Recall that a single operation at most doubles the bit-length of the ciphertext. Hence after any one operation the ciphertext cannot be larger than $2x_\gamma'$ and therefore the sequence of modular reductions involves only small multiples of the x_i', which means that it only adds a small amount of noise.

It is not clear to what extent adding these larger integers to the public key influences the security of the scheme.

Fully homomorphic encryption (FHE) allows a worker to perform implicit additions and multiplications on plaintext values while exclusively manipulating encrypted data. The fully homomorphic scheme proceeds in several steps. First, one constructs a somewhat homomorphic encryption scheme, which only supports a limited number of multiplications: ciphertexts contain some noise that becomes larger with successive homomorphic multiplications, and only ciphertexts whose noise size remains below a certain threshold can be decrypted correctly. The second step is to squash the decryption procedure associated with an arbitrary ciphertext so that it can be expressed as a low-degree polynomial in the secret key bits. Then, the key idea, called bootstrapping, consists of homomorphically evaluating this decryption polynomial on encryptions of the secret key bits, resulting in a different ciphertext associated with the same plaintext, but with possibly reduced noise. This refreshed ciphertext can then be used in subsequent homomorphic operations. By repeatedly refreshing ciphertexts, the number of homomorphic operations becomes unlimited, resulting in a fully homomorphic encryption scheme.

Theorem 3.8 ([6]). *There is a (efficient, explicit) transformation that given a description of a bootstrapped encryption scheme ϵ and a parameter $d = d(\lambda)$ where λ is the security parameter, output a description of another encryption scheme $\epsilon^{(d)}$ such that $\epsilon^{(d)}$ is homomorphic for all circuits of depth up to d.*

3.4.3   Implementation

A fully homomorphic encryption scheme [5] that uses only simple integer arithmetic is described as above. The primary open problem is to improve the efficiency of the scheme, to the extent that it is possible while preserving the hardness of the approximate-GCD problem.

Gentry and Halevi [8] implemented the Gentry's fully homomorphic encryption scheme [6]. The performance can be found in [8].

Coron et al. [3] extended the fully homomorphic encryption scheme over the integers of van Dijk et al. (DGHV) [5] to batch fully homomorphic encryption,

Table 3.1 Parameters of batch DGHV scheme

Instance	λ	ℓ	ρ	η	$\gamma \times 10^{-6}$	τ	Θ	pk size
Toy	42	10	26	988	0.29	188	150	647 kB
Small	52	37	41	1,558	1.6	661	555	13.3 MB
Medium	62	138	56	2,128	8.5	2,410	2,070	304 MB
Large	72	531	71	2,698	39	8,713	7,965	5.6 GB

Table 3.2 Performance of batch DGHV scheme

Instance	KeyGen	Encrypt	Decrypt	Mult	Expand	Recrypt
Toy	0.06 s	0.02 s	0 s	0.003 s	0.007 s	0.11 s
Small	1.74 s	0.23 s	0.02 s	0.025 s	0.08 s	1.10 s
Medium	73 s	3.67 s	0.45 s	0.16 s	1.60 s	11.9 s
Large	3493 s	61 s	9.8 s	0.72 s	28 s	172 s

i.e., to a scheme that supports encrypting and homomorphically processing a vector of plaintext bits as a single ciphertext. They also implemented the batch DGHV scheme, based on a C++ implementation using the GMP library. Tables 3.1 and 3.2 list concrete key sizes and timings for their batch DGHV scheme.

For all security levels, $n = 4$ and $\theta = 15$. In addition, ℓ is the length of the vector for parallel processing.

References

1. Z. Brakerski, C. Gentry, V. Vaikuntanathan, (Leveled) fully homomorphic encryption without bootstrapping, in *Proceedings of the 3rd Innovations in Theoretical Computer Science Conference, ITCS'12*, 2012, pp. 309–325
2. D. Coppersmith, Small solutions to polynomial equations, and low exponent RSA vulnerabilities. J. Cryptol. **10**(4), 233–260 (1997)
3. J.S. Coron, T. Lepoint, M. Tibouchi, Batch fully homomorphic encryption over the integers, in *Proceedings of Advances in Cryptology, EUROCRYPT'13*, 2013, pp. 315–335
4. J.S. Coron, D. Naccache, M. Tibouchi, Public key compression and modulus switching for fully homomorphic encryption over the integers, in *Proceedings of Advances in Cryptology, EUROCRYPT'12*, 2012, pp. 446–464
5. M. van Dijk, C. Gentry, S. Halevi, V. Vaikuntanathan, Fully homomorphic encryption over the integers, in *Proceedings of Advances in Cryptology, EUROCRYPT'10*, 2010, pp. 24–43
6. C. Gentry, Fully homomorphic encryption using ideal lattices, in *Proceedings of STOC'09*, 2009, pp. 169–178
7. C. Gentry, *Fully Homomorphic Encryption Using Ideal Lattices*. PhD thesis, 2009
8. C. Gentry, S. Halevi, Implementing Gentry fully-homomorphic encryption scheme, in *Proceedings of Advances in Cryptology, EUROCRYPT'11*, 2011, pp.129–148
9. S. Halevi, An implementation of homomorphic encryption. http://github.com/shaih/HELib
10. N. Howgrave-Graham, Approximate integer common divisors, in *Proceedings of Cryptology and Latticed, CaLC'01*, 2001, pp. 51–66
11. J.C. Lagarias, The computational complexity of simultaneous diophantine approximation problems. SIAM J. Comput. **14**(1), 196–209 (1985)

12. P.Q. Nguyen, I. Shparlinski, On the insecurity of a server-aided RSA protocol, in *Proceedings of Advances in Cryptology, ASIACRYPT'01*, 2001, pp. 21–35
13. P.Q. Nguyen, J. Stern, Adapting density attacks to low-weight knapsacks, in *Proceedings of Advances in Cryptology, ASIACRYPT'05*, 2005, pp. 41–58
14. R.L. Rivest, A. Shamir, L. Adleman, A method for obtaining digital signatures and public-key cryptosystems. Commun. ACM, **21**(2), 120–126 (1978)
15. N. Smart, F. Vercauteren, Fully homomorphic encryption with relatively small key and ciphertext sizes, in *Proceedings of PKC'10*, 2010, pp. 420–443
16. D. Stehle, R. Steinfeld, Faster fully homomorphic encryption, in *Proceedings of Advances in Cryptology, ASIACRYPT'10*, 2010, pp. 377–394

Chapter 4
Remote End-to-End Voting Scheme

Abstract Recently, remote voting systems have gained popularity and have been used for government elections and referendums in the United Kingdom, Estonia, and Switzerland as well as municipal elections in Canada and party primary elections in the United States and France. Current remote voting schemes assume either the voter's personal computer is trusted or the voter is not physically coerced. In this chapter, we describe a remote end-to-end voting scheme [23], in which the voter's choice remains secret even if the voter's personal computer is infected by malware or the voter is physically controlled by the adversary. Based on homomorphic encryption, the overhead for tallying in such scheme is linear in the number of candidates. Thus, such scheme is practical for elections at a large scale, such as general elections.

4.1 Introduction

Essentially, an end-to-end voting system can be envisioned as a decryption network composed of a collection of election authorities. The network takes as input a collection of encrypted ballots (posted publicly by voters) in one end and outputs in another end the tally of votes (posted publicly by the authorities) with a mathematical proof that the encrypted ballots were decrypted properly and that the votes were unmodified. Informally, an end-to-end voting system achieves *integrity* if any voter can verify that his or her ballot is included unmodified in a collection of ballots, and the public can verify that the collection of ballots produces the correct final tally, and the system keeps *privacy* if no voter can demonstrate how he or she voted to any third party.

So far, there have been two main categories of end-to-end voting schemes—polling station voting schemes and remote voting schemes.

Polling station voting schemes, such as [1,11,17,20,22], build their security on an untappable channel implemented as a private voting booth at a polling place, where a voter can cast his or her ballot in private. Thus, risk of voter coercion and vote buying can be greatly mitigated. These schemes require a voter to vote in person at a polling station on election days. This may not be convenient for those voters who have no access to any polling station on election days, e.g., overseas citizens and military voters.

© Xun Yi, Russell Paulet, Elisa Bertino 2014   67
X. Yi et al., *Homomorphic Encryption and Applications*, SpringerBriefs
in Computer Science, DOI 10.1007/978-3-319-12229-8_4

Remote voting schemes, such as [4, 7, 10, 15], allow people to cast their votes over the Internet, most likely through a Web browser, from home, or possibly any other location where they have Internet access. While voting of this kind is hoped to encourage higher voter turnout and makes accurate accounting for votes easier, it also carries the potential of making abuse easier to perform, especially at a large scale [15]. One challenge to remote voting is how to prevent voter coercion and vote buying because the behavior of a voter casting a ballot remotely can be physically controlled by an adversary. Another challenge is how to ensure the remote personal computer by which a voter casts his or her vote is trusted because malware can endanger integrity of the elections as well as privacy of the voter [16].

The first voting scheme was introduced by Chaum [4], based on a mix network, where a collection of tally authorities take as input a collection of encrypted votes and output a collection of plain votes according to a secret permutation. This scheme allows each voter to make sure his or her vote was counted, while preserving the privacy of the vote as long as at least one tally authority is honest. In order to improve efficiency in tallying, Cohen (Benaloh) and Fischer [8] proposed a voting scheme, based on a homomorphic encryption E, where $E(x)E(y) = E(x + y)$ for any x and y in its domain. The basic idea is for each voter to encrypt his or her vote using a public-key homomorphic encryption function. The encrypted votes are then summed using homomorphic property without decrypting them. Finally, a collection of tallying authorities cooperate to decrypt the final tally. This scheme also preserves the privacy of votes as long as at least one tally authority is honest. In order to provide with unconditional privacy of votes, Fujioka et al. [10] proposed a voting scheme, based on blind signature, where a signer can digitally sign a document without knowing what was signed. The basic idea is that the voter has his or her ballot blindly signed by the voting authority and later publishes the ballot using an anonymous channel. Current voting schemes are based on either mix network, or homomorphic encryption, or blind signature.

The notion of receipt-freeness was first introduced by Benaloh and Tuinstra [2] to model the security of a voting scheme against voter coercion and vote buying. A voting scheme is receipt-freeness if a voter cannot prove to an attacker that he or she voted in a particular manner, even if the voter wishes to do so. Receipt-freeness voting schemes, such as [2, 13, 21], assume the existence of a private voting booth to isolate the voter from the coercer at the moment he or she casts his or her vote. Remote voting schemes are required to be coercion resistant where the voter can be physically controlled by the adversary during voting. A rigorous definition for coercion resistance was given by Juels et al. [15]. This model considers a powerful adversary who can demand coerced voters to vote in a particular manner, abstain from voting, or even disclose their secret keys. A voting scheme is coercion resistant if it is infeasible for the adversary to determine if a coerced voter compiles with the demands. Intuitively, coercion resistance implies receipt-freeness which itself implies privacy.

A coercion-resistant remote voting scheme was demonstrated by Juels et al. [15]. The basic idea is that each voter casts his or her ballot together with a secret credential, both encrypted by the public keys of the tally authorities. After a collection of encrypted ballots are mixed with a mix network such as [12, 14, 18], the validity of ballots (i.e., the validity of credentials) is checked blindly against the voter roll and only valid ballots are decrypted and counted. This scheme does not require an untappable channel for a voter to cast his or her ballot, but instead assumes an untappable channel for a voter to obtain a secret credential from the registrars during registration (potentially using post mail).

Current coercion-resistant remote voting schemes, such as Juels et al.'s scheme [15] and its variants [7], require public-key encryptions on the side of the voter. Thus, they require the voter to trust the personal computer actually casting the ballot on his or her behalf. Considering that the voter's personal computer can be infected by malware that may reveal the voter's preferences or even change the encrypted ballot cast by the voter, Kutylowski and Zagorski [16] recently proposed a remote voting scheme, a combination of paper-based voting schemes Punchscan [5] and ThreeBallot [20]. The basic idea is that a voter makes a complete ballot by laying a ballot and a coding card side by side. Each voter is issued exactly one ballot by the election authority and she or he can get a coding card from any proxy. This scheme preserves privacy of votes if both authorities do not collude. Even if the voter's personal computer is infected by viruses, his or her choice remains secret. This scheme does not allow a voter to prove how he or she voted unless vote casting is physically supervised by an adversary.

Current remote voting schemes assume either the voter's personal computer is trusted to cast a vote or the voter is not physically controlled by the adversary. In this chapter, we describe a remote voting scheme [23], in which the voter's choice remains secret even if the voter's personal computer is infected by malware or the voter is physically controlled by the adversary. The presentation is based on the paper by Yi and Okamoto [23].

The approach by Yi and Okamoto is motivated by the most efficient voting scheme by Hirt and Sako [13] based on homomorphic encryption. The main difference between the approach by Hirt and Sako and the one presented in the chapter is that Hirt and Sako assume the availability of an untappable channel between the voter and the authorities during voting while the approach described in this chapter requires the untappable channel during voter registration only.

Consider an election where the candidates are $\{C_1, C_2, \cdots, C_{n_C}\}$ and the choice for each candidate is either "Yes" or "No"; their basic idea can be described as follows. First of all, a voter $\mathcal{V}_i$ generates a public/private key pair for digital signature scheme on his or her own device. During registration, $\mathcal{V}_i$ presents himself or herself to a registrar's office, where he or she is allowed privately to input n_C references $r_{i,j}$ ($\in \{1, -1\}$) on a trusted entry device (like setting PIN number in a bank branch), which, in turn, encrypts each $g^{r_{i,j}}$ with the public keys of tally authorities according to ElGamal encryption scheme [9] where g is a generator of a cyclic group $\mathbb{G}$ and then posts on a public bulletin board the ciphertexts $\mathsf{E}(g^{r_{i,j}}) = (A_{i,j}, B_{i,j})$ (each corresponds to one candidate C_j) along with the voter's public key. During voting, $\mathcal{V}_i$ posts on the public bulletin board his or her ballot composed of $\beta_j \in \{1, -1\}$ ($j = 1, 2, \cdots, n_C$) and his or her signature on it, where $\beta_j = 1$ if the choice of $\mathcal{V}_i$ is the same as his or her reference $r_{i,j}$ and

$\beta_j = -1$ otherwise. During tallying, the tallying authorities sum $(A_{i,j}{}^{\beta_j}, B_{i,j}{}^{\beta_j})$ for each candidate C_j and then cooperate to decrypt the final tally.

Compared to most of existing remote voting schemes, the scheme described in this chapter has three merits as follows: (1) no encryption is needed during voting and the ballot cast by a voter is "plain"; thus, any voter can verify that his or her ballot is included unmodified; (2) no mix network is needed during tallying and the tallying overhead is linear in the number of candidates; therefore it is practical for elections at a large scale; (3) verifiability remains even if all election authorities are corrupt.

In addition, this scheme allows a voter repeatedly to refresh his or her references remotely after he or she registers and to use refresh references for a new election. Privacy is built on voter registration protected by a untappable channel.

4.2 Remote End-to-End Voting

4.2.1 Participating Parties

Assume that there exists a publicly readable, insert-only bulletin board ($\mathcal{BB}$) on which public information (e.g., public keys, ballots, and final tally) is posted. No one can overwrite or erase existing data on $\mathcal{BB}$. The public (including voters) can read the contents of $\mathcal{BB}$ anytime.

Normally, the remote voting scheme involves three types of participants as follows:

- Registrar ($\mathcal{R}$) authorizes voters for an election by posting each voter's identity and public information on $\mathcal{BB}$.
- Voters ($\mathcal{V}_1, \mathcal{V}_2, \cdots, \mathcal{V}_{nv}$) are the entities participating in the election administrated by $\mathcal{R}$.
- Tallying authorities ($\mathcal{T}_1, \mathcal{T}_2, \cdots, \mathcal{T}_{nT}$) process ballots, jointly count votes, and publish the final tally.

4.2.2 Basic Remote Voting Scheme

We now introduce a basic remote voting scheme, where there is only one candidate, and the choice of the election is either "Yes" or "No."

Setup: The scheme is built on ElGamal (homomorphic and threshold) encryption scheme (ES) [9], the modified ElGamal signature scheme (SS) [19], the non-interactive zero-knowledge reencryption proof (ReencPf) [3, 13], and the non-interactive zero-knowledge equal discrete logarithm proof (EqDlog) [6], over a group $\mathbb{G}$ of a large prime order q with a generator g.

Let the choices of the election be $C = \{1, -1\}$, where $1, -1$ stand for "Yes" and "No," respectively.

Let the list of tallying authorities be $\mathcal{T} = \{\mathcal{T}_1, \mathcal{T}_2, \cdots, \mathcal{T}_{n_T}\}$. Each $\mathcal{T}_i$ randomly chooses a private key $TSK_i = t_i$ from $\mathbb{Z}_q^*$ and computes the public key

$$TPK_i = g^{t_i} \tag{4.1}$$

Let $TSK = \{t_1, t_2, \cdots, t_{n_T}\}$ and $TPK = \{g^{t_1}, g^{t_2}, \cdots, g^{t_{n_T}}\}$. Let h be chosen from a family of collision-resistant hash functions.

At last, the registrar posts $\Omega = \{\mathcal{R}, \mathsf{ES}, \mathsf{SS}, \mathsf{ReencPf}, \mathsf{EqDlog}, (\mathbb{G}, q, g), h, C,$ $\mathcal{T}, TPK, \mathcal{V}\}$ on the public bulletin board $\mathcal{BB}$.

Registration: Before registration, each voter $\mathcal{V}_i$ generates a public/private key pair $(sk_i = x_i, pk_i = g^{x_i})$ for the signature scheme SS on his or her own device and prints out the public key pk_i and the hash value $h(pk_i)$ on paper. The purpose of using a hash function is to facilitate human checking.

To vote, a voter $\mathcal{V}_i$ presents himself or herself to a registrar's office, where $\mathcal{V}_i$ is allowed privately to press Yes or No button on a trusted entry device, which, in turn, encrypts g or g^{-1} accordingly, and then prints out the hash value $h(R_i)$ on a slip, where $R_i = (A_i, B_i)$ is an encryption of either g or g^{-1}. Let

$$r_i = 1, A_i = g^{\gamma_i}, B_i = g \left(\prod_{t=1}^{n_T} TPK_t \right)^{\gamma_i}$$

if press Yes, and let

$$r_i = -1, A_i = g^{\gamma_i}, B_i = g^{-1} \left(\prod_{t=1}^{n_T} TPK_t \right)^{\gamma_i}$$

if press No, where γ_i is randomly chosen by the device from $\mathbb{Z}_q^*$. Therefore, R_i is an encryption of g^{r_i}. The voter $\mathcal{V}_i$ needs to remember his or her reference r_i. Having seen $h(R_i)$ on the slip, the voter $\mathcal{V}_i$ is allowed to confirm his or her choice by pressing "Confirm" or "Cancel" button on the device, like [1, 16].

If $\mathcal{V}_i$ presses "Cancel," the device prints out r_i, γ_i, R_i on the slip for $\mathcal{V}_i$ to check if r_i is his or her choice. In this case, the staff in the registrar's office tears off the slip and provides a handwriting signature on it. $\mathcal{V}_i$ either keeps the slip for anyone later to check or inserts the slip into a locked box placed in the registrar's office for the election inspector with key later to check. Then the registration restarts. Note that anyone can check if R_i on the slip is computed correctly with r_i, γ_i, TPK without the knowledge of private keys of tallying authorities.

If the voter $\mathcal{V}_i$ presses "Confirm," the device scans his or her identity (denoted as $\mathcal{V}_i$ as well) from his or her identity card and his or her public key pk_i from his or her paper and then computes the hash value $h(pk_i)$ and prints out $\mathcal{V}_i, h(pk_i)$ on the slip. The voter needs to check if the hash value $h(pk_i)$ on the slip is the same as that on his or her paper. At last, the device provides non-interactive zero-knowledge reencryption proof P_i (using ReencPf) that R_i is a reencryption of either $(1, g)$ or $(1, g^{-1})$, posts V_i, pk_i, $h(pk_i)$, R_i, $h(R_i)$, P_i on $\mathcal{BB}$, and then erases r_i, γ_i from its memory. The staff tears off the slip with $h(R_i)$, V_i, $h(pk_i)$, provides a handwriting signature on it, and then hands it to the voter.

Let the list of registered voters be $\mathcal{V} = \{V_1, V_2, \cdots, V_{nv}\}$. For each V_i, there is a row

$$(V_i, pk_i, h(pk_i), R_i, h(R_i), P_i)$$

on $\mathcal{BB}$.

Voting: The registrar $\mathcal{R}$ announces the candidate on $\mathcal{BB}$. Each V_i chooses his or her vote v_i from $C = \{1, -1\}$ and determines β_i as follows: If $v_i = r_i$, then $\beta_i = 1$. If $v_i \neq r_i$, then $\beta_i = -1$. Note that V_i remembers his or her reference r_i.

Next, V_i generates a signature on β_i (using $\mathcal{SS}$) as follows:

$$S_i = g^{\delta_i} \tag{4.2}$$

$$T_i = (H(\beta_i, S_i) - S_i x_i)\delta_i^{-1} (mod\ q) \tag{4.3}$$

where δ_i is randomly chosen from $\mathbb{Z}_q^*$, H is a hash function, and x_i is the private key of V_i. Note that a time stamp may be included in the message to be signed to prevent replaying attacks.

Then, V_i constructs a ballot

$$b_i = \{\beta_i, S_i, T_i\}$$

and casts it to $\mathcal{R}$, which, in turn, posts b_i next to V_i on $\mathcal{BB}$ if

$$g^{H(\beta_i, S_i)} = pk_i^{S_i} S_i^{T_i} \tag{4.4}$$

The voter V_i checks if b_i on $\mathcal{BB}$ is the same as that he or she casts.

Tallying: To tally all valid ballots posted on $\mathcal{BB}$, $\mathcal{T}$ performs the following steps:

1. **Combining**: Based on the homomorphic property of ElGamal encryption scheme, all valid ballots $\{b_i\}_{i=1}^{nv}$ on $\mathcal{BB}$ can be combined as follows:

$$X_T = \prod_{i=1}^{nv} A_i^{\beta_i} \tag{4.5}$$

$$Y_T = \prod_{i=1}^{nv} B_i^{\beta_i} \tag{4.6}$$

2. **Decrypting**: Following the threshold ElGamal encryption scheme, each tally authority $\mathcal{T}_i$ computes

$$X_i = X_T^{t_i} \tag{4.7}$$

and posts X_i on $\mathcal{BB}$. With $\{X_i\}_{i=1}^{n_V}$, one can compute

$$Y_T \prod_{i=1}^{n_V} X_i^{-1} = \prod_{i=1}^{n_V} g^{r_i \beta_i} = \prod_{i=1}^{n_V} g^{v_i} = g^{\mathbf{y} - \mathbf{n}},$$

where $\mathbf{y}, \mathbf{n}$ are the numbers of "Yes" and "No" and

$$\mathbf{y} + \mathbf{n} = n_V \tag{4.8}$$

Since n_V is a small number relative to q, $\mathbf{y}$ can be determined from $g^{2\mathbf{y} - n_V} = g^{\mathbf{y} - \mathbf{n}}$ by exhaustively searching $\mathbf{y}$ from 1 to n_V. At last, $\mathcal{T}$ release a tally

$$\mathbf{X} = (\mathbf{y}, \mathbf{n})$$

on $\mathcal{BB}$.

3. **Proving**: $\mathcal{T}_1, \mathcal{T}_2, \cdots, \mathcal{T}_{n_T}$ jointly provide a multiparty non-interactive zero-knowledge proof P (using EqDlog) that

$$\prod_{i=1}^{n_T} TPK_i = g^{\sum_{i=1}^{n_T} t_i} \tag{4.9}$$

$$g^{\mathbf{n} - \mathbf{y}} Y_T = X_T^{\sum_{i=1}^{n_T} t_i} \tag{4.10}$$

have the equal discrete logarithm and then post the proof P next to $\mathbf{X}$ on $\mathcal{BB}$.

Verifying: During registration, each voter $\mathcal{V}_i$ is able to check if his or her public key pk_i and ciphertext R_i are posted on $\mathcal{BB}$ correctly on the basis of hash values $h(pk_i)$ and $h(R_i)$ on his or her registration slip. In addition, $\mathcal{V}_i$ is able to detect if the entry device in the registrar's office cheats by pressing "Cancel" and checking if r_i on the test slip is his or her choice and if R_i on the test slip is computed correctly by himself or herself or with the help of someone later. During voting, each voter $\mathcal{V}_i$ is able to check whether β_i (either 1 or -1) in the ballot $b_i = \{\beta_i, S_i, T_i\}$ posted on $\mathcal{BB}$ is his or her choice even if the computer of $\mathcal{V}_i$ is infected by malware.

During registration, the election inspector is able to detect if the entry device cheats voters by collecting all test slips with the handwriting signatures of the registrar from the test box and checking if all ciphertexts are computed correctly. During voting, the public (including the voters) is able to verify if each R_i is an encryption of either g or g^{-1} based on the non-interactive zero-knowledge proof P_i, and check if each ballot b_i is valid with the signature (S_i, T_i) of $\mathcal{V}_i$. During tallying, the public can check if all valid ballots are combined and decrypted correctly based on the non-interactive zero-knowledge proof P.

Remark. The basic scheme can fit a two-candidate election trivially.

4.2.3 General Remote Voting Scheme

The basic remote voting scheme can be used to build a general remote voting scheme, where there is a list of candidates $\mathbb{C} = \{C_1, C_2, \cdots, C_{nc}\}$, and the choice for each candidate is either "Yes" or "No."

Setup: Same as the basic scheme, the registrar $\mathcal{R}$ posts $\Omega = \{\mathcal{R}, \mathsf{ES}, \mathsf{SS}, \mathsf{ReencPf}, \mathsf{EqDlog}, (\mathbb{G}, q, g), h, C, \mathcal{T}, TPK, \mathcal{V}\}$ on the public bulletin board $\mathcal{BB}$.

Registration: For registration, a voter $\mathcal{V}_i$ presents himself or herself with his or her printed public key pk_i and hash value $h(pk_i)$ to the registrar's office, where $\mathcal{V}_i$ is allowed privately to enter an integer

$$\mathbf{r}_i = a_{i,1} + a_{i,2}2 + \cdots + a_{i,nc}2^{nc-1} \tag{4.11}$$

where $a_{i,j}$ is either 0 or 1, into a trusted entry device, which, in turn, encrypts a series of g and g^{-1} according to $a_{i,j}$. The ciphertext $R_{i,j} = (A_{i,j}, B_{i,j})$ and

$$(A_{i,j}, B_{i,j}) = \begin{cases} (g^{\gamma_{i,j}}, g(\prod_{t=1}^{n_T} TPK_t)^{\gamma_{i,j}}) & \text{if } a_{i,j} = 0 \\ (g^{\gamma_{i,j}}, g^{-1}(\prod_{t=1}^{n_T} TPK_t)^{\gamma_{i,j}}) & \text{if } a_{i,j} = 1 \end{cases}$$

where $\gamma_{i,j}$ is randomly chosen by the device from $\mathbb{Z}_q^*$. Then the device prints out the hash value $h(\mathbb{R}_i)$ on a slip, where $\mathbb{R}_i = \{R_{i,j}\}_{j=1}^{nc}$. The voter $\mathcal{V}_i$ needs to remember his or her reference $\mathbf{r}_i$ (like a PIN number). If the number of candidates is large, $\mathcal{V}_i$ may write down $\mathbf{r}_i$ on a note privately.

Having seen $h(\mathbb{R}_i)$ on the slip, $\mathcal{V}_i$ decides whether to confirm $\mathbf{r}_i$. If not, the device prints out $\mathbf{r}_i$, $\{\gamma_{i,j}\}_{j=1}^{nc}$ and $\mathbb{R}_i$ on the slip. In this case, the staff in the registrar's office tears off the slip and provides a handwriting signature on it. $\mathcal{V}_i$ either keeps the slip for anyone later to check or inserts the slip into a locked box placed in the registrar's office for the election inspector with key later to check. Then the registration restarts. Otherwise, the device scans the identity $\mathcal{V}_i$ and the public key pk_i and prints out $\mathcal{V}_i, h(pk_i)$ on the slip for $\mathcal{V}_i$ to check. At last, the device provides a non-interactive zero-knowledge reencryption proofs $\mathbb{P}_i$ (using $\mathsf{ReencPf}$) that each ciphertext in $\mathbb{R}_i$ is a reencryption of either $(1, g)$ or $(1, g^{-1})$, and erases $\mathbf{r}_i, a_{i,j}, \gamma_{i,j}$ from its memory, and posts

$$\mathcal{V}_i, pk_i, h(pk_i), \mathbb{R}_i, h(\mathbb{R}_i), \mathbb{P}_i$$

on $\mathcal{BB}$. The staff tears off the slip with $h(\mathbb{R}_i), \mathcal{V}_i, h(pk_i)$, provides a handwriting signature on it, and then hands it to the voter.

Voting: The registrar $\mathcal{R}$ announces the list of candidates $\mathbb{C} = \{C_1, C_2, \cdots, C_{nc}\}$ on $\mathcal{BB}$.

For each candidate C_j ($j = 1, 2, \cdots, n_C$), a voter $\mathcal{V}_i$ chooses his or her vote $v_{i,j}$ from $\{1, -1\}$ and determines $\beta_{i,j}$ as follows: If $v_{i,j} = (-1)^{a_{i,j}}$, then $\beta_{i,j} = 1$. If $v_{i,j} \neq (-1)^{a_{i,j}}$, then $\beta_{i,j} = -1$. Note that $\mathcal{V}_i$ remembers his or her reference $\mathbf{r}_i = a_{i,1} + a_{i,2}2 + \cdots + a_{i,n_C}2^{n_C - 1}$.

Next, $\mathcal{V}_i$ generates a signature on $\{\beta_{i,1}, \beta_{i,2}, \cdots, \beta_{i,n_C}\}$ as follows:

$$S_i = g^{\delta_i} \tag{4.12}$$

$$T_i = (H(\beta_{i,1}, \beta_{i,2}, \cdots, \beta_{i,n_C}, S_i) - S_i x_i)\delta_i^{-1} (mod\ q) \tag{4.13}$$

where δ_i is randomly chosen from $\mathbb{Z}_q^*$ and x_i is the private key of $\mathcal{V}_i$.

Then, $\mathcal{V}_i$ constructs a ballot

$$b_i = \{\{\beta_{i,j}\}_{j=1}^{n_C}, S_i, T_i\}$$

and casts it to $\mathcal{R}$, which, in turn, posts b_i next to $\mathcal{V}_i$ on $\mathcal{BB}$ if

$$g^{H(\beta_{i,1}, \beta_{i,2}, \cdots, \beta_{i,n_C}, S_i)} = pk_i^{S_i} S_i^{T_i} \tag{4.14}$$

The voter $\mathcal{V}_i$ checks if b_i on $\mathcal{BB}$ is the same as that he or she casts.

Tallying: To tally all valid ballots posted on $\mathcal{BB}$ for each candidate C_j ($j = 1, 2, \cdots, n_C$), $\mathcal{T}$ performs the following steps:

1. **Combining**: $\mathcal{T}$ combines all valid ballots on $\mathcal{BB}$ for the candidate C_j as follows:

$$X_{T,j} = \prod_{i=1}^{n_V} A_{i,j}{}^{\beta_{i,j}} \tag{4.15}$$

$$Y_{T,j} = \prod_{i=1}^{n_V} B_{i,j}{}^{\beta_{i,j}} \tag{4.16}$$

2. **Decrypting**: Each tally authority $\mathcal{T}_i$ computes $X_{i,j} = X_{T,j}{}^{t_i}$ and posts $X_{i,j}$ on $\mathcal{BB}$. By $\{X_{i,j}\}_{i=1}^{n_V}$, one can compute

$$Y_{T,j} \cdot \prod_{i=1}^{n_V} X_{i,j}{}^{-1} = \prod_{i=1}^{n_V} g^{\beta_{i,j}(-1)^{a_{i,j}}} = \prod_{i=1}^{n_V} g^{v_{i,j}} = g^{\mathbf{y}_j - \mathbf{n}_j}$$

where $\mathbf{y}_j, \mathbf{n}_j$ are the numbers of "Yes" and "No" for the candidate C_j and $\mathbf{y}_j + \mathbf{n}_j = n_V$. Then $\mathbf{y}_j$ can be determined from $g^{2\mathbf{y}_j - n_V} = g^{\mathbf{y}_j - \mathbf{n}_j}$ by exhaustively searching $\mathbf{y}_j$ from 1 to n_V. At last, $\mathcal{T}$ release a tally

$$\mathbf{X}_j = (\mathbf{y}_j, \mathbf{n}_j)$$

for the candidate C_j on $\mathcal{BB}$.

3. **Proving**: Tallying authorities $T_1, T_2, \cdots, T_{n_T}$ jointly provide a multiparty non-interactive zero-knowledge proof P_{C_j} (using EqDlog) that

$$\prod_{i=1}^{n_T} TPK_i = g^{\sum_{i=1}^{n_T} t_i} \tag{4.17}$$

$$g^{\mathbf{n}_j - \mathbf{y}_j} Y_{T.j} = X_{T.j}^{\sum_{i=1}^{n_T} t_i} \tag{4.18}$$

have the equal discrete logarithm and then post the proof P_{C_j} next to $\mathbf{X}_j$ on $\mathcal{BB}$.

Verifying: Same as the basic voting scheme.

Remark. The general scheme can fit an m out of n selection election (where $m < n$), in which m candidates are elected from n candidates $C_1, C_2, \cdots, C_n$, as long as we rank $\mathbf{y}_i - \mathbf{n}_i$ ($i = 1, 2, \cdots, n_C$) after tallying. In addition, the general scheme can be extended to a ranked election. For example, considering a ranked election with 4 candidates C_1, C_2, C_3, C_4, a voter can rank them by 4 preferences $(+, +), (+, -), (-, +)$, and $(-, -)$. To implement this, each voter presets two ciphertexts $R_{i,j,1}, R_{i,j,2}$ on $\mathcal{BB}$ for each candidate C_j. After voting, two columns of ciphertexts for C_j are tallied, respectively, and the tallying result for C_j can be $2(\mathbf{y}_{j,1} - \mathbf{n}_{j,1}) + (\mathbf{y}_{j,2} - \mathbf{n}_{j,2})$, where $(\mathbf{y}_{j,k}, \mathbf{n}_{j,k})$ is the tallying result of the kth column of the ciphertexts for C_j.

4.2.4   Voter Reference Refresh

In the basic and general remote voting schemes, the reference of a voter can be used for one election only. For a new election, the voter may go to the registrar's office to reset his or her reference as the registration described above or refresh his or her reference online as follows.

For the basic scheme, when the voter V_i refreshes his or her reference r_i ($\in \{1, -1\}$), whose ciphertext on $\mathcal{BB}$ is $R_i = (A_i, B_i)$, he or she randomly chooses μ_i from $\{1, -1\}$ while his or her computer randomly chooses ρ_i from $\mathbb{Z}_q^*$ and computes

$$R_i' = (A_i', B_i') = (g^{\rho_i} A_i^{\mu_i}, \left(\prod_{t=1}^{n_T} TPK_t\right)^{\rho_i} B_i^{\mu_i}) \tag{4.19}$$

where R_i' is an encryption of $g^{r_i \mu_i}$ and the refresh reference $r_i' = r_i \mu_i$. Next the computer of V_i provides a non-interactive zero-knowledge reencryption proof P_i' that R_i' is a reencryption of either (A_i, B_i) or (A_i^{-1}, B_i^{-1}) and generates a signature on R_i' as follows:

$$S_i' = g^{\delta_i'} \tag{4.20}$$

$$T_i' = (H(R_i', S_i') - S_i' x_i)\delta_i'^{-1} (mod \ q) \tag{4.21}$$

where δ_i' is randomly chosen from $\mathbb{Z}_q^*$ and x_i is the private key of $\mathcal{V}_i$. At last, $\mathcal{V}_i$ posts

$$(R_i', S_i', T_i', P_i')$$

next to $\mathcal{V}_i$ on $\mathcal{BB}$.

Remark. If an adversary coerces a voter $\mathcal{V}_i$ to compute R_i' with μ_i and ρ_i chosen by himself or herself, he or she is uncertain of the refresh reference $r_i' = r_i \mu_i$ because he or she is uncertain of the original reference r_i. In case the registrar obtains r_i, γ_i during the registration of $\mathcal{V}_i$, the registrar is uncertain of the refresh reference $r_i' = r_i \mu_i$ because he or she is uncertain of μ_i.

For the general scheme, when the voter $\mathcal{V}_i$ refreshes his or her reference $\mathbf{r}_i$ ($= a_{i,1} + a_{i,2}2 + \cdots + \cdots a_{i,nc}2^{nc-1}$), where $a_{i,j} \in \{0, 1\}$), he or she randomly chooses $\mu_{i,1}, \cdots, \mu_{i,nc}$ from $\{1, -1\}$, while his or her computer chooses random numbers $\rho_{i,1}, \rho_{i,2}, \cdots, \rho_{i,nc}$ from $\mathbb{Z}_q^*$ and computes

$$\mathbb{R}_i' = \{(A_{i,j}', B_{i,j}')\}_{j=1}^{nc} = \left\{\left(g^{\rho_{i,j}} A_i^{\mu_{i,j}}, \left(\prod_{t=1}^{nT} TPK_t\right)^{\rho_{i,j}} B_i^{\mu_{i,j}}\right)\right\}_{j=1}^{nc} \tag{4.22}$$

where $\mathbb{R}_i'$ is the set of the encryptions of

$$(g^{\mu_{i,1}(-1)^{a_{i,1}}}, g^{\mu_{i,2}(-1)^{a_{i,2}}}, \cdots, g^{\mu_{i,nc}(-1)^{a_{i,nc}}})$$

and the refresh reference

$$\mathbf{r}_i' = a_{i,1}' + a_{i,2}'2 + \cdots + a_{i,nc}'2^{nc-1} \tag{4.23}$$

where

$$a_{i,j}' = \frac{1 - \mu_{i,j}(-1)^{a_{i,j}}}{2} \tag{4.24}$$

Next $\mathcal{V}_i$ provides a non-interactive zero-knowledge reencryption proof $\mathbb{P}_i'$ that each $(A_{i,j}', B_{i,j}')$ in $\mathbb{R}_i'$ is a reencryption of either $(A_{i,j}, B_{i,j})$ or $(A_{i,j}^{-1}, B_{i,j}^{-1})$ and generates a signature on $\mathbb{R}_i'$ as follows:

$$S_i' = g^{\delta_i'} \tag{4.25}$$

$$T_i' = (H(\mathbb{R}_i', S_i') - S_i' x_i)\delta_i'^{-1} (mod \ q) \tag{4.26}$$

where δ_i' is randomly chosen from $\mathbb{Z}_q^*$ and x_i is the private key of $\mathcal{V}_i$. At last, $\mathcal{V}_i$ posts $(\mathbb{R}_i', S_i', T_i', \mathbb{P}_i)$ next to $\mathcal{V}_i$ on $\mathcal{BB}$.

Remark. As a voter $\mathcal{V}_i$ is able to test if the entry device in the registrar's office is cheating during the registration, $\mathcal{V}_i$ is able to test if his or her computer is cheating during voter reference refresh by sending test data to the election inspector by post.

4.3   Conclusion and Discussion

In this chapter, we have described an Internet voting system [23]. While the overhead for tallying in Juels et al.'s remote voting system [15] is quadratic in the number of voters, the overhead for tallying in the Internet voting system described in this chapter is only $O(n_V)$ which is linear in the number of voters. Therefore, the system is practical for elections at a large scale, such as general elections. In addition, Juels et al.'s remote voting system [15] is not verifiable in the sense that an adversary, who has corrupted all tallying authorities, is able to forge valid ballots without being detected. The Internet voting system in this chapter overcomes this drawback. Even if the adversary corrupts all election authorities, the adversary is unable to forge any valid ballot in the system. At last, a voter in the Internet voting system does not need to encrypt his or her ballot during voting. The ballot is in a form of plaintext. Therefore, even if the voter's personal computer is infected by malware, any modification on the voter's ballot can be detected by the voter. Furthermore, his or her vote choice remains secret because his or her final vote is a combination of his or her ballot and his or her reference which is encrypted during registration and posted on the bulletin board.

References

1. J. Benaloh, Ballot casting assurance via voter-initiated poll station auditing, in *Proceedings of Electronic Voting Technology Workshop (EVT'07)*, 2007
2. J. Benaloh, D. Tuinstra. Receipt-free secret-ballot elections (extended abstract), in *Proceedings of 26th ACM STOC'94*, 1994, pp. 544–553
3. M. Blum, A.D. Santis, S. Micali, G. Persiano, Non-interactive zero-knowledge. SIAM J. Comput. **6**, 1084–1118 (1991)
4. D. Chaum, Untraceable electronic mail, return addresses, and digital pseudonyms. Commun. ACM **24**(2), 84–88 (1981)
5. D. Chaum, *Punchscan*, 2005, http://www.punchscan.org.
6. D. Chaum, T.P. Pedersen, Wallet databases with observers, in *Proceedings of CRYPTO'92*, 1992, pp. 89–105
7. M.R. Clarkson, S. Chong, A.C. Myers, Civitas: A secure remote voting system, in *Proceedings of SP'08*, 2008, pp. 354–368
8. J.D. Cohen (Benaloh) M.J. Fischer, A robust and verifiable cryptographically secure election scheme, in *Proceedings of FOCS'85*, pp. 372–382, 1985.

9. T. ElGamal, A public key cryptosystem and a signature scheme based on discrete logarithms. IEEE Trans. Infor. Theo. **31**, 469–472 (1985)
10. A. Fujioka, T. Okamoto, K. Ohta, A practical secret voting scheme for large scale elections, in *Proceedings of AUSCRYPT'92*, pp. 244–251, 1992.
11. R.W. Gardner, S. Garera, A.D. Rubin, Coercion resistant end-to-end voting, *Proceedings of FC'09*, 2009, pp. 344–361
12. P. Golle, M. Jakobsson, A. Juels P. Syverson, Universal re-encryption for mixnets, in *Proceedings of CT-RSA'04*, 2004, pp. 163–178
13. M. Hirt, K. Sako, Efficient receipt-free voting based on homomorphic encryption, in *Proceedings of EUROCRYPT'00*, 2000, pp. 539–556
14. M. Jakobsson, A. Juels, R. Rivest, Making mix nets robust for electronic voting by randomized partial checking, in *Proceedings of USENIX'02*, 2002, pp. 339–353
15. A. Juels, D. Catalano, M. Jakobsson, Coercion-resistant electronic election, in *Proceedings of WPES'05*, 2005, pp. 61–70
16. M. Kutylowski, F. Zagorski, Scratch, click & vote: E2E voting over the Internet. NIST End-to-End Voting System Workshop, 2009.
17. T. Moran, M. Naor, Split-ballot voting: everlasting privacy with distributed trust, in *Proceedimgs of CCS'07*, 2007, pp. 246–255
18. A. Neff, A verifiable secret shuffle and its application to e-voting, in *Proceedings of CCS'01*, 2001, pp. 116–125
19. D. Pointcheval, J. Stern. Security proofs for signature schemes, in *Proceedings of EUROCRYPT'96*, 1996, pp. 387–398
20. R.L. Rivest, W.D. Smith, Three voting protocols: Threeballot, VAV, and twin, in *Proceedings of Electronic Voting Technology Workshop (EVT'07)*, 2007, pp. 16–16
21. K. Sako, J. Kilian, Receipt-free mix-type voting scheme-a practical solution to the implementation of a voting booth, in *Proceedings of EUROCRYPT'95*, 1995, pp. 393–403
22. V. Teague, K. Ramchen, L. Naish. Coercion-resistant tallying for STV voting, in *Proceedings of Electronic Voting Technology Workshop (EVT'08)*, 2008
23. X. Yi, E. Okamoto, Practical Internet voting system. J. Netw. Comput. Appl. **36**(1), 378–387 (2013)

Chapter 5
Nearest Neighbor Queries with Location Privacy

Abstract In mobile communication, spatial queries pose a serious threat to user location privacy because the location of a query may reveal sensitive information about the mobile user. In this chapter, we consider k nearest neighbor (kNN) queries where the mobile user queries the location-based service (LBS) provider about k nearest points of interest (POIs) on the basis of his or her current location. We described a solution given by Yi et al. [22] for the mobile user to preserve his or her location privacy in kNN queries. The solution is built on the Paillier public-key cryptosystem [11] and can provide both location privacy and data privacy. In particular, the solution allows the mobile user to retrieve one type of POIs, for example, k nearest car parks, without revealing to the LBS provider what type of points is retrieved. For a cloaking region with $n \times n$ cells and m types of points, the total communication complexity for the mobile user to retrieve a type of k nearest POIs is $O(n + m)$ while the computation complexities of the mobile user and the LBS provider are $O(n + m)$ and $O(n^2m)$, respectively. Compared with existing solutions for kNN queries with location privacy, these solutions are more efficient.

5.1 Introduction

The embedding of positioning capabilities (e.g., GPS) in mobile devices facilitates the emergence of location-based services (LBSs), which are considered as the next "killer application" in the wireless data market. LBS allows clients to query a service provider (such as Google or Bing Maps) in a ubiquitous manner, in order to retrieve detailed information about points of interest (POIs) in their vicinity (e.g., restaurants, hospitals, etc.).

The LBS provider processes spatial queries on the basis of the location of the mobile user. Location information collected from mobile users, knowingly and unknowingly, can reveal far more than just a user's latitude and longitude. Knowing where a mobile user is can mean knowing what he/she is doing: attending a religious service or a support meeting, visiting a doctor's office, shopping for an engagement ring, carrying out non-work-related activities in office, or spending an evening at the corner bar. It might reveal that he or she is interviewing for a new job or "out" him or her as a participant at a gun rally or a peace protest. It can mean knowing with

© Xun Yi, Russell Paulet, Elisa Bertino 2014

X. Yi et al., *Homomorphic Encryption and Applications*, SpringerBriefs in Computer Science, DOI 10.1007/978-3-319-12229-8_5

whom he/she spends time and how often. When location data are aggregated it can reveal his/her regular habits and routines—and when he or she deviates from them.

A 2010 survey conducted for Microsoft in the United Kingdom, Germany, Japan, the United States, and Canada found that 94 % of consumers who had used LBSs considered them valuable, but the same survey found that 52 % were concerned about potential loss of privacy.[1]

In this chapter, we consider k nearest neighbor (kNN) queries where the mobile user queries the LBS provider about k nearest POIs. In general, the mobile user needs to submit his or her location to the LBS provider which then finds out and returns to the user the k nearest POIs by comparing the distances between the mobile user's location and POIs nearby. This reveals the mobile user's location to the LBS provider.

There have been numerous techniques that can provide a certain degree of location privacy. These techniques mainly include

- Information access control [10, 23];
- Mix zone [2];
- k-Anonymity [1, 3, 9];
- "Dummy" locations [8, 16, 21];
- Geographic data transformation [6, 7, 19, 20];
- Private information retrieval (PIR) [4, 5, 12–14].

Localtion-based service queries based on access control, mix zone, and k-anonymity require the service provider or the middleware that maintains all user locations. They are vulnerable to misbehavior of the third party. They offer little protection when the service provider/middleware is owned by an untrusted party. There have been private data inadvertently disclosed over the Internet in the past.

k-Anonymity is initially used for identity privacy protection. It is generally inadequate for location privacy protections, where the notion of distance between locations is important (unlike distances between identities). The effect of LBS queries based on k-anonymity depends heavily on the distribution and density of the mobile users, which, however, are beyond the control of the location privacy technique.

Location-based service queries based on dummy locations require the mobile user randomly to choose a set of fake locations, to send the fake locations to the LBS, and to receive the false reports from the LBS over the mobile network. This incurs both computation and communication overhead in mobile devices. For the purpose of efficiency, the mobile user may choose fewer fake locations, but the LBS provider can restrict the user in a small subspace of the total domain, leading to weak privacy.

Location-based service queries based on geographic data transformation are prone to access pattern attacks [18] because the same query always returns the same encoded results. For example, the LBS may observe the frequencies of the returned

---

[1]http://www.microsoft.com/security/resources/research.aspx#LBS.

ciphertexts. Having knowledge about the context of the database, it can match the most popular plaintext POI with the most frequently returned ciphertext and, thus, unravel information about the query.

Location-based service queries based on PIR provide strong cryptographic guarantees, but are often computationally and communicationally expensive. To improve efficiency, trusted hardware was employed to perform PIR for LBS queries [12]. This technique is built on hardware-aided PIR [17], which assumes that a trusted third party (TTP) initializes the system by setting the secret key and the permutation of the database. Like LBS queries based on access control, mix zone, and k-anonymity, this technique is vulnerable to misbehavior of the third party.

It is a challenge to give practical solutions for kNN queries with location privacy on the basis of PIR.

In this chapter, we describe some solutions for kNN queries by Yi et al. [22] on the basis of PIR with the Paillier public-key cryptosystem [11]. Yi et al.'s work has three main contributions as follows:

- Current PIR-based LBS queries [4, 5, 13, 14] usually require two stages. In the first stage, the mobile user retrieves the index of his or her location from the LBS provider. In the second stage, the mobile user retrieves the POIs according to the index from the LBS provider. To simplify the process, Yi et al. give a solution for kNN queries which needs one stage only, i.e., the mobile user sends his or her location (encrypted) to the LBS provider and receives the k nearest POIs (encrypted) from the LBS provider.
- Current PIR-based LBS queries only allow the mobile user to find out k nearest POIs regardless of the type of POIs. For the first time, Yi et al. take into account the type of POIs in kNN queries and give a solution for the mobile user to find out k nearest PIOs of the same type without revealing to LBS provider what type of POIs he or she is interested in.
- Current PIR-based LBS queries all need to fix a cloaking region based on which the LBS provider generates the responses to the mobile user's queries. If the cloaking region is large, the LBS queries are inefficient. If the cloaking region is small, the LBS queries have weak privacy. Yi et al. give a solution for the mobile user to specify a large public cloaking region but let the LBS provider generate the responses actually based on a small private cloaking region repeatedly.

For a cloaking region with $n \times n$ cells and m types of points, assume that the mobile user wishes to retrieve a type of k nearest POIs at his or her location, the total communication complexity is $O(n + m)$ while the computation complexities of the mobile user and the LBS provider are $O(n + m)$ and $O(n^2 m)$, respectively. Compared with previous solutions for kNN queries with location privacy, this solution is more efficient.

5.2 Private k Nearest Neighbor Queries

5.2.1 Security Model

The security model considers an LBS scenario in mobile environments, as shown in
Fig. 5.1, where there exist the mobile user, the LBS provider, the base station and
satellites, each playing a different role.

- The mobile user sends location-based queries to the LBS provider and receives
  LBS from the provider.
- The LBS provider provides LBSs to the mobile user.
- The base station bridges the mobile communications between the mobile user
  and the LBS provider.
- Satellites provide the location information to the mobile user.

We assume that the mobile user can acquire his or her location from satellites
anonymously, and the base station and the LBS provider do not collude to compro-
mise the user location privacy or there exists an anonymous channel such as Tor[2]
for the mobile user to send queries to and receive services from the LBS provider.
The model focuses on user location privacy protection against the LBS provider
and a kNN query protocol (where k is fixed) is composed of three algorithms as
follows:

1. Query Generation (QG): Taking as input a cloaking region CR with $n \times n$ cells
   and m distinct types of POIs, the location (i, j) of the mobile user, and the type
   t of POIs; the mobile user outputs a query Q (containing CR) and a secret s,
   denoted as $(Q, s) = \text{QG}(CR, n, m, (i, j), t)$.
2. Response Generation (RG): Taking as input the query Q and the location-based
   database D of POIs; the LBS provider outputs a response R, denoted as $R = \text{RG}(Q, D)$.

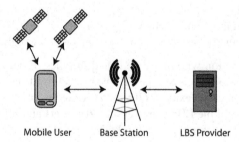

Mobile User   Base Station   LBS Provider

Fig. 5.1 Location-based service

---

[2]https://www.torproject.org/.

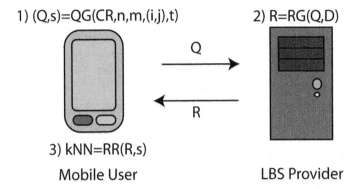

1) (Q,s)=QG(CR,n,m,(i,j),t)   2) R=RG(Q,D)

Q

R

3) kNN=RR(R,s)

Mobile User   LBS Provider

Fig. 5.2 Private kNN query

3. Response Retrieval (RR): Taking as input the response R and the secret s of the mobile user; the mobile user outputs k nearest POIs of the type t, denoted as $kNN = \mathsf{RR}(R, s)$.

A private kNN query protocol can be illustrated in Fig. 5.2 and is correct if $kNN = \mathsf{RR}(R, s)$ outputs k nearest POIs of the type t corresponding to the cell at (i, j), where $(Q, s) = \mathsf{QG}(CR, n, m, (i, j), t)$ and $R = \mathsf{RG}(Q, D)$.

The security of a private kNN query protocol involves data privacy and location privacy. Intuitively, the LBS provider S wishes to release only the k nearest POIs of one type to the mobile user $\mathcal{U}$ each time when the user sends a kNN query. Meanwhile, the mobile user $\mathcal{U}$ does not wish to reveal to the LBS provider his or her location (i, j) and the type t of POIs he or she is interested in.

Formally, data privacy can be defined with a game as follows.

Given a user location (i, j) where $1 \leq i, j \leq n$ and one type t of POIs, consider the following game between an adversary (the user) $\mathcal{A}$ and a challenger $\mathcal{C}$. The game consists of the following steps:

1. The adversary chooses any two distinct cloaking regions CR_1 and CR_2 with $n \times n$ cells such that k nearest POIs of the type t in the cell (i, j) are same. The adversary generates a query Q to retrieve the k nearest POIs of the type t in the cell (i, j) and sends Q, CR_1, CR_2 to the challenger $\mathcal{C}$.
2. The challenger $\mathcal{C}$ chooses a random bit $b \in \{0, 1\}$, and runs the Response Generation algorithm RG to obtain $R_b = \mathsf{RG}(Q(CR_b), D)$, and then sends R_b back to $\mathcal{A}$.
3. The adversary $\mathcal{A}$ can experiment with the code of R_b in an arbitrary non-black-box way. If the adversary can retrieve the k nearest POIs of the type t in the cell (i, j) from R_b, he or she outputs $b' \in \{0, 1\}$.

The adversary wins the game if $b' = b$ and loses otherwise. We define the adversary $\mathcal{A}$'s advantage in this game to be

$$\mathsf{Adv}_{\mathcal{A}}(k) = |\Pr(b' = b) - 1/2|,$$

where k is the security parameter.

Definition 5.1 (Data Privacy Definition). In a kNN query protocol, the LBS provider has data privacy if for any probabilistic polynomial time (PPT) adversary $\mathcal{A}$, we have that $\mathsf{Adv}_{\mathcal{A}}(k)$ is a negligible function, where the probability is taken over coin-tosses of the challenger and the adversary.

Remark. Data privacy ensures that the response distributions on the user's view are computationally indistinguishable for any two cloaking regions CR_1 and CR_2 such that the k nearest POIs of the type t in the cell (i, j) in the two cloaking regions are the same. This means that a computationally bounded user does not receive information about more than one cell in the cloaking region CR.

Next, we formally define location privacy with a game as follows.

Given a cloaking region CR with $n \times n$ cells and m types of POIs, consider the following game between an adversary (the LBS provider) $\mathcal{A}$ and a challenger $\mathcal{C}$. The game consists of the following steps:

1. The adversary $\mathcal{A}$ chooses two distinct tuples (i_0, j_0, t_0) and (i_1, j_1, t_1), where (i_b, j_b) represents the cell and t_b stands for the type of POIs, from the cloaking region CR and sends them to the challenger $\mathcal{C}$.
2. The challenger $\mathcal{C}$ chooses a random bit $b \in \{0, 1\}$ and executes the Query Generation (QG) to obtain $(Q_b, s) = \mathsf{QG}(CR, n, m, (i_b, j_b), t_b)$ and then sends Q_b back to the adversary $\mathcal{A}$.
3. The adversary $\mathcal{A}$ can experiment with the code of Q_b in an arbitrary non-black-box way and finally outputs a bit $b' \in \{0, 1\}$.

The adversary wins the game if $b' = b$ and loses otherwise. We define the adversary $\mathcal{A}$'s advantage in this game to be

$$\mathsf{Adv}_{\mathcal{A}}(k) = |\Pr(b' = b) - 1/2|$$

where k is the security parameter.

Definition 5.2 (Location Privacy Definition). In a kNN query protocol, the user has location privacy if for any probabilistic polynomial time (PPT) adversary $\mathcal{A}$, we have that $\mathsf{Adv}_{\mathcal{A}}(k)$ is a negligible function, where the probability is taken over coin-tosses of the challenger and the adversary.

Remark. Location privacy ensures that the server cannot determine the location of the mobile user in the cloaking region CR and the type of POIs with the kNN query from the mobile user.

Based on the model, we describe some constructions of private kNN query protocol by Yi et al. [22] which allows the mobile user to find k nearest POIs from a cloaking region. These solutions are built on the Paillier homomorphic encryption scheme [11] and the Rabin encryption scheme [15].

5.2.2 Private kNN Queries Without Data Privacy

First of all, we describe a basic construction of kNN query protocol without considering data privacy of the LBS provider. We assume that there is only one type of POIs and so we ignore the type of POIs and t in the model in this case.

Initially, the LBS provider divides the location-based database D (a geographic map) into cells with the same size, for example, 1 km width and 1 km length. Based on the center of each cell, the LBS provider collects k nearest POIs, $P_1, P_2, \cdots, P_k$ as shown in Fig. 5.3 and each point is represented by a tuple (x, y), where x and y are the latitude and longitude of the point, respectively. For each cell (i, j), the LBS provider keeps k nearest POIs, represented as a stream of bits, denoted as an integer $d_{i,j}$. We assume $M = max(d_{i,j})$, i.e., the longest record.

Remark. Because the LBS provider collects k nearest POIs according to the center of each cell (i.e., the red points shown in Fig. 5.3), the LBS provider responds the same k nearest POIs to the two mobile users within the same cell no matter where the two mobile users are in the cell. For the mobile user located near the border of two cells, he or she may query two cells or even four cells around his or her location and then find out k nearest POIs among the query responses. The purpose of this method is to avoid privately comparing distances, which is hard to do without revealing the location of the user.

We assume that the mobile user $\mathcal{U}$ wishes to find k nearest POIs around his or her location. To do so, the user $\mathcal{U}$ chooses a cloaking region CR with $n \times n$ cells, where $\mathcal{U}$ is located in the cell (i, j), and runs the kNN query protocol with the LBS provider $\mathcal{S}$, composed of three algorithms, Query Generation (QG), Response Generation (RG), and Response Retrieval (RR), as described in Algorithms 1–3.

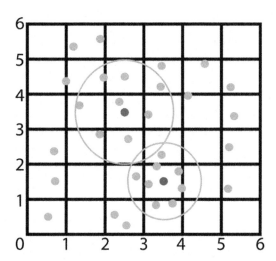

Fig. 5.3 k Nearest POIs for cells

Algorithm 1 Query Generation (user)

Input: $CR, n, (i, j)$
Output: Q, s
 1: Randomly choose two large primes p, q such that $N = pq > M$.
 2: Let $sk = \{p, q\}$ and $pk = \{g, N\}$, where g is chosen from $\mathbb{Z}_{N^2}$ and its order is a nonzero multiple of N.
 3: For each $\ell \in \{1, 2, \cdots, n\}$, pick a random integer $r_\ell \in \mathbb{Z}_{N^2}^*$, compute

$$c_\ell = \begin{cases} Encrypt(1, pk) = g^1 r_\ell^N (mod\ N^2) & \text{if } \ell = i \\ Encrypt(0, pk) = g^0 r_\ell^N (mod\ N^2) & \text{otherwise} \end{cases}$$

where the encryption algorithm is described in the Paillier cryptosystem.
 4: Let $Q = \{CR, n, c_1, c_2, \cdots, c_n, pk\}$, $s = sk$.
 5: **return** Q, s

Algorithm 2 Response Generation RG (server)

Input: $D, Q = \{CR, n, c_1, c_2, \cdots, c_n, (g, N)\}$
Output: $R = \{C_1, C_2, \cdots, C_n\}$
 1: Based on CR and n, compute $R = \{C_1, C_2, \cdots, C_n\}$ where for $\gamma = 1, 2, \cdots, n$,

$$C_\gamma = \prod_{\ell=1}^{n} c_\ell^{d_{\ell,\gamma}} (mod\ N^2)$$

 2: **return** R

Algorithm 3 Response Retrieval RR (user)

Input: $R = \{C_1, C_2, \cdots, C_n\}$, $sk = s$
Output: d
 1: Compute

$$d = Decrypt(C_j, sk),$$

where the decryption algorithm is described in the Paillier cryptosystem.
 2: **return** d

Remark. The CR may be specified by the coordinates (x, y) of an origin point and the order n of a square grid. The cell which contains the origin point is labeled as $(1,1)$. The CR covers the square grid from the cell $(1,1)$ to the cell (n, n).

Remark. In Algorithm 3, when the mobile user receives the response, he or she can ignore C_ℓ ($\ell \neq j$) and receive C_j only because only C_j contains the information about the k nearest POIs in the cell (i, j).

Theorem 5.3 (Correctness). *The kNN query protocol without considering data privacy of the LBS provider (Algorithms 1–3) is correct. In other words, for any cloaking region CR with $n \times n$ and the index i, j of a cell ($1 \leq i, j \leq n$), we have*

Algorithm 4 Query Generation (user)

Input: $CR, n, (i, j)$
Output: Q, s

1: Randomly choose two large primes p, q such that $N = pq > M$.
2: Let $sk = \{p, q\}$ and $pk = \{g, N\}$, where g is chosen from $\mathbb{Z}_{N^2}$ and its order is a nonzero multiple of N.
3: For each $\ell \in \{1, 2, \cdots, n\}$, pick a random integer $r_\ell \in \mathbb{Z}_{N^2}^*$, compute

$$c_\ell = \begin{cases} Encrypt(1, pk) = g^1 r_\ell^N (mod\ N^2) & \text{if } \ell = i \\ Encrypt(0, pk) = g^0 r_\ell^N (mod\ N^2) & \text{otherwise} \end{cases}$$

4: Pick a random integer $r \in \mathbb{Z}_{N^2}^*$, compute

$$c = Encrypt(j, pk) = g^j r^N (mod\ N^2)$$

5: Let $Q = \{CR, n, c_1, c_2, \cdots, c_n, c, pk\}, s = sk$.
6: **return** Q, s

$$d_{i,j} = \mathsf{RR}(R, s),$$

where $d_{i,j}$ stands for k nearest POIs for the cell (i, j), $(Q, s) = \mathsf{QG}(CR, n, (i, j))$, $R = \mathsf{RG}(D, Q)$.

Proof. Based on Algorithms 1–3, we have

$$C_j = \prod_{\ell=1}^{n} c_\ell^{d_{\ell,j}} = g^{d_{i,j}} \left(\prod_{\ell=1}^{n} r_\ell^{d_{\ell,j}} \right)^N (mod\ N^2),$$

which is a Paillier encryption of $d_{i,j}$. Therefore, we have $d_{i,j} = Decrypt(C_j, sk)$ $= \mathsf{RR}(R, sk)$ and the theorem is proved.   □

5.2.3   Private kNN Queries with Data Privacy

In the kNN query protocol without considering data privacy of the LBS provider, $C_\gamma = g^{d_{i,\gamma}} (\prod_{\ell=1}^{n} r_\ell^{d_{\ell,j}})^N (mod\ N^2)$ and thus the mobile user is able to obtain the k nearest POIs in cells (i, γ) for $\gamma = 1, 2, \cdots, n$. Therefore, it does not have data privacy which requires that the mobile user retrieves the k nearest POIs for one cell only per query.

Now we describe a construction of the kNN query protocol by Yi et al. [22], composed of Algorithms 4–6, which provides data privacy for the LBS provider.

Theorem 5.4 (Correctness). *The kNN query protocol with data privacy (Algorithms 4–6) is correct. In other words, for any cloaking region CR with $n \times n$ and the index i, j of a cell $(1 \leq i, j \leq n)$,*

Algorithm 5 Response Generation RG (server)

Input: $D, Q = \{CR, n, c_1, c_2, \cdots, c_n, c, (g, N)\}$
Output: $R = \{C_1, C_2, \cdots, C_n\}$
  1: Based on CR and n, compute $R = \{C_1, C_2, \cdots, C_n\}$ where for $\gamma = 1, 2, \cdots, n$,

$$C_\gamma = (c/g^\gamma)^{w_\gamma} \prod_{\ell=1}^{n} c_\ell^{d_{\ell,\gamma}^2} \ (mod \ N^2),$$

  where w_γ is randomly chosen from $\mathbb{Z}_N^*$.
  2: **return** R

Algorithm 6 Response Retrieval RR (user)

Input: $R = \{C_1, C_2, \cdots, C_n\}, sk = s$
Output: d
  1: Compute

$$C_j' = PaillierDecrypt(C_j, sk),$$

  where the decryption algorithm is described in the Paillier cryptosystem.
  2: Compute

$$d = RabinDecrypt(C_j', sk),$$

  where the decryption algorithm is described in the Rabin cryptosystem.
  3: **return** d

$$d_{i,j} = RR(R, s)$$

holds, where $d_{i,j}$ stands for k nearest POIs, $(Q, s) = QG(CR, n, (i, j))$, $R = RG(D, Q)$.

Proof. Based on Algorithms 4–6, we have

$$C_j = (c/g^j)^{w_j} \prod_{\ell=1}^{n} c_\ell^{d_{\ell,j}^2} \ (mod \ N^2)$$

$$= g^{d_{i,j}^2} \left(r^{w_j} \prod_{\ell=1}^{n} r_\ell^{d_{\ell,j}^2} \right)^N \ (mod \ N^2),$$

which is a Paillier encryption of $d_{i,j}^2 \ (mod \ N)$. Therefore, we have

$$C_j' = Paillier \ Decrypt(C_j, sk) = d_{i,j}^2 \ (mod \ N)$$

which is the Rabin encryption of $d_{i,j}$. At last, we have

$$d_{i,j} = RabinDecrypt(C_j, sk) = \mathsf{RR}(R, s)$$

and the theorem is proved.   □

Remark. For any γ,

$$C_\gamma = g^{d_{i,\gamma}^2} g^{(j-\gamma)w_\gamma} \left(r^{w_\gamma} \prod_{\ell=1}^{n} r_\ell^{d_{\ell,j}^2} \right)^N$$

When $\gamma \neq j$, C_γ is not a Paillier encryption of $d_{i,\gamma}^2$ because of $g^{(j-\gamma)w_\gamma}$. This means that the mobile user cannot obtain k nearest POIs for the cell (i, γ) when $\gamma \neq j$. In addition, we use the Rabin encryption $d_{i,j}^2$ instead of $d_{i,j}$ in the Response Generation to prevent the mobile user from retrieving the nearest POIs for the cell (ℓ, j) when $\ell \neq i$. If we encode $d_{i,j}$ rather than $d_{i,j}^2$, a malicious user may retrieve a linear equation of $d_{1,j}, d_{2,j}, \cdots, d_{n,j}$ by including more than one encryption of 1 in the list of $c_1, c_2, \cdots, c_n$. The linear relation may disclose more than one $d_{i,j}$ to the user. By Rabin encryption, the user can only retrieve a nonlinear equation of $d_{1,j}, d_{2,j}, \cdots, d_{n,j}$ if there are more than one encryption of 1 in the list of $c_1, c_2, \cdots, c_n$. From the nonlinear equation, it is hard to retrieve any $d_{i,j}$.

5.2.4 Private kNN Queries Based on POI Type

Now we take the POI type in kNN query into account. Slightly different from the initialization phase in the kNN query protocol without data privacy, based on the center of each cell, the LBS provider collects k nearest POIs, $P_1, P_2, \cdots, P_k$ and each point is represented by a tuple (x, y, t), where x and y are the latitude and longitude of the point, respectively, and t is the type of the points. Examples of POI types includes:

- Churches, schools
- Post offices, shops, postboxes, telephone boxes
- Pubs
- Car parks
- Speed cameras
- Tourist attractions

We assume that POI types are coded into $1, 2, \cdots, m$ which is published to the public. For each cell (i, j) and each POI type t, the LBS keeps k nearest POIs of type t, represented by a stream of bits, denoted as an integer $d_{i,j,t}$. We assume $M = max(d_{i,j,t})$.

Assume that the mobile user $\mathcal{U}$ located in the cell (i, j) wishes to find k nearest POIs of the type t; the kNN query protocol based on POI type is composed of Algorithms 7–9.

Algorithm 7 Query Generation (user)

Input: $CR, n, m, (i, j), t$
Output: Q, s
1: Randomly choose two large primes p_1, q_1 such that $N_1 = p_1 q_1 > M$.
2: Randomly choose two large primes p_2, q_2 such that $N_2 = p_2 q_2$, where $N_1^2 < N_2 < N_1^4$.
3: Let $sk_1 = \{p_1, q_1\}, sk_2 = \{p_2, q_2\}, pk_1 = \{g_1, N_1\}, pk_2 = \{g_2, N_2\}$, where g_1 is chosen from $\mathbb{Z}_{N_1^2}$ and its order is a nonzero multiple of N_1 and g_2 is chosen from $\mathbb{Z}_{N_2^2}$ and its order is a nonzero multiple of N_2.
4: For each $\ell \in \{1, 2, \cdots, m\}$, pick a random integer $r_\ell \in \mathbb{Z}_{N_1^2}^*$, compute

$$c_\ell = \begin{cases} E(1, pk_1) = g_1^1 r_\ell^{N_1} (mod \ N_1^2) & \text{if } \ell = t \\ E(0, pk_1) = g_1^0 r_\ell^{N_1} (mod \ N_1^2) & \text{otherwise} \end{cases}$$

5: For each $\ell \in \{1, 2, \cdots, n\}$, pick a random integer $r_\ell' \in \mathbb{Z}_{N_2^2}^*$, compute

$$c_\ell' = \begin{cases} E(1, pk_2) = g_2^1 r_\ell'^{N_2} (mod \ N_2^2) & \text{if } \ell = i \\ E(0, pk_2) = g_2^0 r_\ell'^{N_2} (mod \ N_2^2) & \text{otherwise} \end{cases}$$

6: Pick a random integer $r \in \mathbb{Z}_{N_2^2}^*$, compute

$$c = E(j, pk_2) = g_2^j r^{N_2} (mod \ N_2^2)$$

7: Let $Q = \{CR, n, m, c_1, c_2, \cdots, c_m, c_1', c_2', \cdots, c_n', c, pk_1, pk_2\}, s = \{sk_1, sk_2\}$.
8: **return** Q, s

Algorithm 8 Response Generation RG (server)

Input: $D, Q = \{CR, m, n, c_1, c_2, \cdots, c_m, c_1', c_2', \cdots, c_n', c, pk_1, pk_2\}$
Output: $R = \{C_1, C_2, \cdots, C_n\}$
1: Based on CR and m, for each cell (α, β) in CR, compute

$$C_{\alpha, \beta} = \prod_{\ell=1}^{m} c_\ell^{d_{\alpha,\beta,\ell}^2} (mod \ N_1^2)$$

2: Based on CR and n, compute $R = \{C_1, C_2, \cdots, C_n\}$, where for $\beta \in \{1, 2, \cdots, n\}$,

$$C_\beta = (c/g^\beta)^{w_\beta} \prod_{\alpha=1}^{n} c_\alpha'^{C_{\alpha,\beta}^2} (mod \ N_2^2),$$

where w_β is randomly chosen from $\mathbb{Z}_{N_2}^*$
3: **return** R

Theorem 5.5 (Correctness). *The kNN query protocol based on POI type (Algorithms 7–9) is correct. In other words, for any cloaking region CR with $n \times n$ and m types of POIs, and the index i, j of a cell $(1 \leq i, j \leq n)$ and a type t of POIs, we have*

Algorithm 9 Response Retrieval RR (user)

Input: $R = \{C_1, C_2, \cdots, C_n\}, sk$
Output: d
1: Compute

$$C'_j = PaillierDecrypt(C_j, sk_2).$$

where the decryption algorithm is described in the Paillier cryptosystem.
2: Compute

$$C''_j = RabinDecrypt(C'_j, sk_2).$$

where the decryption algorithm is described in the Rabin cryptosystem.
3: Compute

$$C'''_j = PaillierDecrypt(C''_j, sk_1).$$

4: Compute

$$d = RabinDecrypt(C'''_j, sk_1).$$

5: **return** d

$$d_{i,j,t} = \mathsf{RR}(R, sk)$$

holds, where $d_{i,j,t}$ stands for k nearest POIs of the type t in the cell (i, j), and $(Q, sk) = \mathsf{QG}(CR, n, m, (i, j), t)$, $R = \mathsf{RG}(D, Q)$.

Proof. Following the proof of Theorem 2, we can prove that

$$C''_j = C_{i,j} = \prod_{\ell=1}^{m} c_\ell^{d_{i,j,\ell}^2} (mod\ N_1{}^2).$$

In fact, $C_{i,j} = g^{d_{i,j,t}^2} (\prod_{\ell=1}^{m} r_\ell^{d_{i,j,\ell}^2})^{N_1} (mod\ N_1^2)$ which is a Paillier encryption of $d_{i,j,t}^2 (mod\ N_1)$. Therefore, we have

$$C'''_j = d_{i,j,t}^2 (mod\ N_1),$$

which is the Rabin encryption of $d_{i,j,t}$. At last, we have $d_{i,j,t} = RabinDecrypt$ $(C'''_j, sk_1) = RR(R, s)$ and the theorem is proved. $\square$

Algorithm 10 Private cloaking region request (user)

Input: CR, Δ, i, j
Output: Q, s
1: Randomly choose two large primes p_1, q_1 such that $N_1 = p_1 q_1 > M$.
2: Randomly choose two large primes p_2, q_2 such that $N_2 = p_2 q_2$, where $N_1^2 < N_2 < N_1^4$.
3: Let $sk_1 = \{p_1, q_1\}, sk_2 = \{p_2, q_2\}, pk_1 = \{g_1, N_1\}, pk_2 = \{g_2, N_2\}$, where g_1 is chosen from $\mathbb{Z}_{N_1^2}$ and its order is a nonzero multiple of N_1 and g_2 is chosen from $\mathbb{Z}_{N_2^2}$ and its order is a nonzero multiple of N_2.
4: For each $\ell \in \{1, 2, \cdots, \Delta\}$, pick a random integer $r_\ell \in \mathbb{Z}_{N_1^2}^*$, compute

$$c_\ell = \begin{cases} E(1, pk_1) = g_1^1 r_\ell^{N_1} \pmod{N_1^2} & \text{if } \ell = i \\ E(0, pk_1) = g_1^0 r_\ell^{N_1} \pmod{N_1^2} & \text{otherwise} \end{cases}$$

5: For each $\ell \in \{1, 2, \cdots, \Delta\}$, pick a random integer $r'_\ell \in \mathbb{Z}_{N_2^2}^*$, compute

$$c'_\ell = \begin{cases} E(1, pk_2) = g_2^1 r_\ell'^{N_2} \pmod{N_2^2} & \text{if } \ell = j \\ E(0, pk_2) = g_2^0 r_\ell'^{N_2} \pmod{N_2^2} & \text{otherwise} \end{cases}$$

6: Let $Q = \{CR, \Delta, c_1, c_2, \cdots, c_\Delta, c'_1, c'_2, \cdots, c'_\Delta, pk_1, pk_2\}, s = \{sk_1, sk_2\}$.
7: **return** Q, s

5.2.5 Private Cloaking Region

In the kNN query protocols, the mobile user needs to specify a cloaking region CR in his or her query Q. If the CR is too large, the kNN query will be inefficient. However, if the CR is too small, the kNN query has weak location privacy.

To facilitate the kNN query protocols, we describe a solution by Yi et al. [22] for the mobile user to specify a (small) private cloaking region (encrypted) in a (big) public cloaking region. After that, the mobile user and the LBS provider can run the kNN query protocols over the private cloaking region repeatedly.

Assume that the public cloaking region CR contains $\Delta \times \Delta$ small cloaking regions $CR_{\alpha,\beta}$ ($\alpha = 1, 2, \cdots, \Delta, \beta = 1, 2, \cdots, \Delta$). Without loss of generality, we assume that each small $CR_{\alpha,\beta}$ contains λ data elements, $d_{\alpha,\beta,\gamma}$ for $\gamma = 1, 2, \cdots, \lambda$, although the small $CR_{\alpha,\beta}$ can be further divided into $n \times n$ cells later.

Assume that the mobile user wishes to specify a private cloaking region $CR_{i,j}$ (encrypted); the private cloaking region protocol is composed of Algorithm 10, by which the mobile user generates a request for private cloaking region, and Algorithm 11, by which the LBS provider generates the private cloaking region (encrypted) for the mobile user.

Before we describe the private cloaking region generation algorithm, we introduce a notation as follows:

$$c_i^{CR_{\alpha,\beta}} = (c_i^{d_{\alpha,\beta,1}}, c_i^{d_{\alpha,\beta,2}}, \cdots, c_i^{d_{\alpha,\beta,\lambda}})$$

Algorithm 11 Private cloaking region generation (server)

Input: $D, Q = \{CR, \Delta, c_1, c_2, \cdots, c_\Delta, c'_1, c'_2, \cdots, c'_\Delta, pk_1, pk_2\}$
Output: R
1: Based on CR and Δ, CR is divided into small cloaking regions $CR_{\alpha,\beta}$ where $1 \le \alpha, \beta \le \Delta$.
2: For $\beta = 1, 2, \cdots, \Delta$, compute

$$CR_\beta = c_1{}^{CR_{1,\beta}} c_2{}^{CR_{2,\beta}} \cdots c_\Delta{}^{CR_{\Delta,\beta}}.$$

3: Compute

$$R = c'_1{}^{CR_1} c'_2{}^{CR_2} \cdots c'_\Delta{}^{CR_\Delta}.$$

4: **return** R

---

and

$$(c_i^{d_{\alpha,\beta,1}}, c_i^{d_{\alpha,\beta,2}}, \cdots, c_i^{d_{\alpha,\beta,\lambda}})(c_j^{d_{\alpha',\beta',1}}, c_j^{d_{\alpha',\beta',2}}, \cdots, c_j^{d_{\alpha',\beta',\lambda}})$$

$$= (c_i^{d_{\alpha,\beta,1}} c_j^{d_{\alpha',\beta',1}}, c_i^{d_{\alpha,\beta,2}} c_j^{d_{\alpha',\beta',2}}, \cdots, c_i^{d_{\alpha,\beta,\lambda}} c_j^{d_{\alpha',\beta',\lambda}})$$

In Algorithm 11, the output R (i.e., the encrypted private cloaking region) contains λ data elements.

Theorem 5.6. *In Algorithms 10 and 11, R is the encryption of private cloaking region $CR_{i,j}$.*

Proof. In Algorithms 10 and 11, assume that $CR_\beta = (C_{\beta,1}, C_{\beta,2}, \cdots, C_{\beta,\lambda})$ and $R = (C_1, C_2, \cdots, C_\lambda)$; then for $\gamma = 1, 2, \cdots, \lambda$, we have

$$C_{\beta,\gamma} = \prod_{\ell=1}^{\Delta} c_\ell^{d_{\ell,\beta,\gamma}} = g_1{}^{d_{i,\beta,\gamma}}(r_{\beta,\gamma})^{N_1}(mod\ N_1),$$

which is a Paillier encryption of $d_{i,\beta,\gamma}$ with g_1, N_1. In addition, for $\gamma = 1, 2, \cdots, \lambda$, we have

$$C_\gamma = \prod_{\ell=1}^{\Delta} c'_\ell{}^{C_{\ell,\gamma}} = g_2{}^{C_{j,\gamma}}(r_\gamma)^{N_2}(mod\ N_2),$$

which is a Paillier encryption of $C_{j,\gamma}$ with g_2, N_2. This means *PaillierDecrypt* $= C_{j,\gamma}$ and *PaillierDecrypt*$(C_{j,\gamma}, sk_1) = d_{i,j,\gamma}$ for $\gamma = 1, 2, \cdots, \lambda$ and the theorem is proved. □

Remark. If the LBS provider has sufficient storage, it can keep the private cloaking region (PCR) for the time being. The PCR is encrypted and only the mobile user can decrypt. The LBS provider still knows the POI types of data elements in the PCR, but it has no idea where PCR is located. Therefore, the mobile user does not need to hide his or her location within the PCR in his or her query and only needs to embed the POI type t in his or her query in the same way as Algorithm 1. In addition, the user can repeatedly query the different cells in the PCR.

5.3 Performance Analysis

Now we analyze the performance of the three kNN query protocols and the private cloaking region protocol by Yi et al. [22]. In the performance analysis, we consider the computation of modular exponentiations (exp.) and ignore the computation of modular multiplications and squares because the latter is much cheaper than the former. We also ignore the process of key generation because it can be precomputed.

5.3.1 Protocol Performance

In the kNN query protocol without data privacy (Algorithms 1–3), the mobile user needs to compute n Paillier encryptions (about n exp.) in Algorithm 1 and 1 Paillier decryption (about 2 exp.) in Algorithm 3. So the total computation complexity of the mobile user is about $O(n)$ exp. In Algorithm 2, the LBS provider needs to compute n^2 exp. and the total computation complexity of the LBS provider is $O(n^2)$ exp. In addition, the communication complexity is $2n \log_2 N$ bits.

In the kNN query protocol with data privacy (Algorithms 4–6), the mobile user needs to compute $n + 1$ Paillier encryptions (about n exp.) in Algorithm 4 and 1 Paillier decryption and 1 Rabin decryption (about 3 exp.) in Algorithm 6. So the total comp. complexity of the user is about $O(n)$ exp. In Algorithm 5, the LBS provider needs to compute $(2 + n)n$ exp. and the total comp. complexity of the LBS provider is $O(n^2)$ exp. In addition, the comm. complexity is $2n \log_2 N$ bits.

In the kNN query protocol based on POI type (Algorithms 7–9), the mobile user needs to compute $n + m + 1$ Paillier encryptions (about $n + m$ exp.) in Algorithm 7 and 2 Paillier decryption and 2 Rabin decryption (about 6 exp.) in Algorithm 9. So the total computation complexity of the mobile user is about $O(2n)$ exp. In Algorithm 8, the LBS provider needs to compute $mn^2 + (n + 2)n$ exp. and the total computation complexity of the LBS provider is $O(mn^2)$ exp. In addition, the communication complexity is $(2n + m) \log_2 N$ bits.

Table 5.1 shows the performance of the above three protocols.

In addition, in the private cloaking region protocol (Algorithms 10–11), the mobile user needs to compute 2Δ Paillier encryptions (about 2Δ exp.) in Algorithm 10 while the LBS provider needs to compute $\lambda\Delta^2$ exp., and the communication complexity is $2\Delta \log_2 N$. After generation of the private cloaking region, the mobile user can repeatedly query it with $O(1)$ (without POI type) or $O(m)$ (with POI type) computation and communication complexities.

Table 5.1 Performance of the kNN query protocols

Component	Algorithms 1–3	Algorithms 4–6	Algorithms 7–9
User comp.	$O(n)$	$O(n)$	$O(n+m)$
Server comp.	$O(n^2)$	$O(n^2)$	$O(mn^2)$
Comm.	$2n \log_2 N$	$2n \log_2 N$	$(2n+m) \log_2 N$

Table 5.2 Performance comparison (stage 1/stage 2)

Component	Ghinita et al.	Paulet et al.	Proposed protocol
User comp.	$O(n^2)/O(n)$	$O(1)$ / generate G, g, q and solve discrete log	$O(n)$
Server comp.	$O(n^2)/O(n^2)$	$O(n)/O(n^2)$	$O(n^2)$
Comm.	$n^2 \log_2 N/2n \log_2 N$	$2n \log_2 N/O(1)$	$2n \log_2 N$

5.3.2 Performance Comparison

We now compare the kNN query protocol with data privacy with PIR-based LBS query protocols [4, 5, 13, 14] in Table 5.2. All these protocols do not consider POI type in their queries. We assume the cloaking region has $n \times n$ cells.

From Table 5.2, we can see that the Ghinita et al. and Paulet et al. protocols both have two stages while the protocol has one stage only. The performance of the protocol is better than the Ghinita et al. protocol in terms of user and server computation complexities and communication complexity. In addition, the Paulet et al. protocol and the protocol have almost the same server computation and communication complexities. The mobile user in the protocol needs to compute much less than the Paulet et al. protocol. In stage 2, the Paulet et al. protocol needs to generate a group G, a generator g, and a prime q for each query and compute a discrete logarithm $c_i = \log_h h_e$. This process takes more time than computing n exp.

5.4 Conclusion and Discussion

In this chapter, we have described the private kNN solution of Yi et al. [22]. Their solution is composed of three private kNN query protocols and one private cloaking region protocol. To analyze the security of the solutions, Yi et al. defined a security model for private kNN queries and performed security analysis on their solution in [22]. The security analysis has shown that the solutions ensure both location privacy in the sense that the user does not reveal any information about his or her location to the LBS provider and data privacy in the sense that the LBS provider releases to the user only k nearest POIs per query. The performance analysis has shown that their protocols are more efficient than the past solutions.

References

1. B. Bamba, L. Liu, P. Pesti, T. Wang, Supporting anonymous location queries in mobile environments with PrivacyGrid, in *Proceedings of the 17th International Conference on World Wide Web, WWW'08*, 2008, pp. 237–246
2. A.R. Beresford, F. Stajano, Location privacy in pervasive computing. IEEE Pervasive Comput. **2**(1), 46–55 (2003)
3. C.Y. Chow, M.F. Mokbel, X. Liu, A peer-to-peer spatial cloaking algorithm for anonymous location-based services, in *Proceedings of the 14th Annual International Symposium on Advances in Geographic Information Systems, ACM GIS'06*, 2006, pp. 171–178
4. G. Ghinita, P. Kalnis, S. Skiadopoulos, PRIVE: Anonymous location-based queries in distributed mobile systems, in *Proceedings of the 16th International Conference on World Wide Web, WWW'07*, 2007, pp. 371–380
5. G. Ghinita, P. Kalnis, A. Khoshgozaran, C. Shahabi, K.-L. Tan, Private queries in location-based services: anonymizers are not necessary, in *Proceedings of International Conference on Management of Data, SIGMOD'08*, 2008, pp. 121–132
6. H. Hu, J. Xu, C. Ren, B. Choi, Processing private queries over untrusted data cloud through privacy homomorphism, in *Proceedings of IEEE 27th International Conference on Data Engineering, ICDE'11*, 2011, pp. 601–612
7. A. Khoshgozaran, C. Shahabi, Blind evaluation of nearest neighbor queries using space transformation to preserve location privacy, in *Proceedings of Advances in Spatial and Temporal Databases, SSTD'07*, 2007, pp. 239–257
8. H. Kido, Y. Yanagisawa, T. Satoh, An anonymous communication technique using dummies for location-based services, in *Proceedings of International Conference on Pervasive Services, ICPS'05*, 2005, pp. 88–97
9. M.F. Mokbel, C.-Y. Chow, W.G. Aref, The new casper: query processing for location services without compromising privacy, in *Proceedings of the 32nd International Conference on Very Large Data Bases, VLDB'06*, 2006, pp. 763–774
10. G. Myles, A. Friday, N. Davies, Preserving privacy in environments with location-based applications. IEEE Pervasive Comput. **2**(1), 56–64 (2003)
11. P. Paillier, Public key cryptosystems based on composite degree residue classes, in *Proceedings of Advances in Cryptology, EUROCRYPT'99*, 1999, pp. 223–238
12. S. Papadopoulos, S. Bakiras, D. Papadias, Nearest neighbor search with strong location privacy, in *Proceedings of the VLDB'10*, 2010, pp. 619–629
13. R. Paulet, M. Golam Kaosar, X. Yi, E. Bertino, Privacy-preserving and content-protecting location based queries, in *Proceedings of IEEE 28th International Conference on Data Engineering ICDE'12*, 2012, pp. 44–53
14. R. Paulet, M. Golam Kaosar, X. Yi, E. Bertino, Privacy-preserving and content-protecting location based queries. IEEE Trans. Knowl. Data Eng. **26**(5), 1200–1210 (2014)
15. M. Rabin, Digitalized signatures and public-key functions as intractable as factorization. (Massachusetts Institute of Technology, Cambridge, 1979)
16. P. Shankar, V. Ganapathy, L. Iftode, Privately querying location-based services with sybilquery, in *Proceedings of the 11th International Conference on Ubiquitous Computing, Ubicomp'09*, 2009, pp. 31–40
17. S. Wang, X. Ding, R.H. Deng, F. Bao, Private information retrieval using trusted hardware, in *Proceedings of Computer Security, ESORICS'06*, 2006, pp. 49–64
18. P. Williams, R. Sion, Usable PIR, in *Proceedings of 15th Annual Network and Distributed System Security Symposium, NDSS'08*, 2008
19. W.K. Wong, D.W. Cheung, B. Kao, N. Mamoulis, Secure kNN computation on encrypted databases, in *Proceedings of International Conference on Management of Data, SIGMOD'09*, 2009, pp. 139–152
20. B. Yao, F. Li, X. Xiao, Secure nearest neighbor revisited, in *Proceedings of IEEE 29th International Conference on Data Engineering, ICDE'13*, 2013, pp. 733–744

21. M.L. Yiu, C. Jensen, X. Huang, H. Lu, SpaceTwist: Managing the trade-offs among location privacy, query performance, and query accuracy in mobile systems, in *Proceedings of IEEE 24th International Conference on Data Engineering, ICDE'08*, 2008, pp. 366–375
22. X. Yi, R. Paulet, E. Bertino, V. Varadharajan, Practical k nearest neighbor queries with location privacy, in *Proceedings of IEEE 30th International Conference on Data Engineering, ICDE'14*, 2014, pp. 640–651
23. M. Youssef, V. Atluri, N.R. Adam, Preserving mobile customer privacy: An access control system for moving objects and custom profiles, in *Proceedings of the 6th MDM'05*, 2005, pp. 67–76

Chapter 6
Private Searching on Streaming Data

Abstract Private searching on streaming data is a process to dispatch to a public
server a program, which searches streaming sources of data without revealing
searching criteria and then sends back a buffer containing the findings. From an
Abelian group homomorphic encryption, the searching criteria can be constructed
by only simple combinations of keywords, e.g., disjunction of keywords. The recent
breakthrough in fully homomorphic encryption has allowed one to construct
arbitrary searching criteria theoretically. In this chapter, we consider a new private
query suggested by Yi et al. [23], which searches for documents from streaming
data on the basis of keyword frequency, such that the frequency of a keyword is
required to be higher or lower than a given threshold. This form of query can
help us in finding more relevant documents. Based on the state-of-the-art fully
homomorphic encryption techniques, we describe disjunctive, conjunctive, and
complement constructions for private threshold queries based on keyword frequency
given by Yi et al. [23]. Combining the basic constructions, we also describe
their generic construction for arbitrary private threshold queries based on keyword
frequency.

6.1 Introduction

The problem of private searching on streaming data was first introduced by
Ostrovsky and Skeith [15]. It was motivated by one of the tasks of the intelligence
community, that is, how to collect potentially useful information from huge volumes
of streaming data flowing through a public server. However, that data which is
potentially useful and raises a red flag is often classified and satisfies secret search
criteria. The challenge is thus how to keep the search criteria classified even if the
program residing in the public server falls into the adversary's hands. This problem
has many applications for the purpose of intelligence gathering. For example, in
airports one can use this technique to find if any of hundreds of passenger lists has
a name from a possible list of terrorists and, if so, to find his/hers itinerary without
revealing the secret terrorists' list.

The first solution for private searching on streaming data was given by Ostrovsky
and Skeith [15, 16]. It was built on the concept of public-key program obfuscation,
where an obfuscator compiles a given program f from a complexity class $\mathcal{C}$ into

© Xun Yi, Russell Paulet, Elisa Bertino 2014   101
X. Yi et al., *Homomorphic Encryption and Applications*, SpringerBriefs
in Computer Science, DOI 10.1007/978-3-319-12229-8_6

a pair of algorithms (F, Dec), such that $Dec(F(x)) = f(x)$ for any input x and it is impossible to distinguish for any polynomial time adversary which f from $\mathcal{C}$ was used to produce a given code for F. The basic idea can be briefly described as follows.

Assume that the public dictionary of potential keywords is $D = \{w_1, w_2, \cdots, w_{|D|}\}$. To search for documents containing one or more of classified keywords $K = \{k_1, k_2, \cdots, k_{|K|}\} \subset D$, the client generates a public/private key pair of a public-key cryptosystem and constructs a program F, composed of an encrypted dictionary $\mathcal{E}(D)$ from K and a buffer $\mathbb{B}$ which will store matching documents. Then the client dispatches the program F to a public server, where F filters streaming documents and stores the encryptions of matching documents in the buffer $\mathbb{B}$. After the buffer $\mathbb{B}$ returns, the client decrypts the buffer and retrieves the matching documents. Because both the keywords and the buffer are encrypted, the search criteria are kept classified to the public.

On the basis of this idea, several solutions for private searching on streaming data have been proposed in literature as follows:

1. Ostrovsky and Skeith [15, 16] gave two solutions for private searching on streaming data. One is based on the Paillier cryptosystem [18] and allows to search for documents satisfying a disjunctive condition $k_1 \vee k_2 \vee \cdots \vee k_{|K|}$, i.e., containing one or more classified keywords. Another is based on the Boneh et al. cryptosystem [3] and can search for documents satisfying $(k_{11} \vee k_{12} \vee \cdots \vee k_{1|K_1|}) \wedge (k_{21} \vee k_{22} \vee \cdots \vee k_{2|K_2|})$, an AND of two sets of keywords.
2. Bethencourt, Song, and Water [1,2] also gave a solution to search for documents satisfying a condition $k_1 \vee k_2 \vee \cdots \vee k_{|K|}$. Like the idea of [17], an encrypted dictionary is used. However, rather than using one large buffer and attempting to avoid collisions like [15], Bethencourt et al. stored the matching documents in three buffers and retrieved them by solving linear systems.
3. Yi et al. [24] proposed a solution to search for documents containing more than t out of n keywords, so-called (t, n) threshold searching, without increasing the dictionary size. The solution is built on the state-of-the-art fully homomorphic encryption (FHE) technique and the buffer keeps at most m matching documents without collisions. Searching for documents containing one or more classified keywords like [1,2,15,16] can be achieved by $(1, n)$ threshold searching.

The existing solutions for private searching on streaming data have not considered keyword frequency, the number of times that keyword is used in a document. Search engines like Google, Yahoo, and AltaVista display results based on secret algorithms. Although we do not know the equations, we believe that these are based mainly on keyword frequency and link popularity.

In this chapter, we describe protocols [23] for a new private query, which searches for documents from streaming data based on keyword frequency, such that a number of times that a keyword appears in a matching document is required to be higher or lower than a given threshold. For example, find documents containing keywords $k_1, k_2, \cdots, k_n$ such that the frequency of the keyword k_i ($i = 1, 2, \cdots, n$) in the document is higher (or lower) than t_i. The protocol takes the lower case into account because terms that appear too frequently are often not very useful as they may not allow one to retrieve a small subset of documents from the streaming data.

This form of query can help one in finding more relevant documents, but it cannot be implemented with traditional homomorphic encryption schemes. Based on fully homomorphic encryption, disjunctive, conjunctive, and complement constructions have been given by Yi et al. [23] for private threshold queries based on keyword frequency: (1) The disjunctive construction allows one to search for documents satisfying a condition such as $(f(k_1) \geq t_1) \vee (f(k_2) \geq t_2) \vee \cdots \vee (f(k_n) \geq t_n)$, where $f(k_i)$ denotes the frequency of the keyword k_i and t_i is a given threshold. (2) The conjunctive construction allows to search for documents satisfying a condition such as $(f(k_1) \geq t_1) \wedge (f(k_2) \geq t_2) \wedge \cdots \wedge (f(k_n) \geq t_n)$. (3) There are two complement constructions. The disjunctive complement construction allows one to search for documents satisfying a condition such as $(f(k_{i_1}) \geq t_{i_1}) \vee \cdots \vee (f(k_{i_{n_1}}) \geq t_{i_{n_1}}) \vee \neg(f(k_{j_1}) \geq t_{j_1}) \vee \cdots \vee \neg(f(k_{j_{n_2}}) \geq t_{j_{n_2}})$, i.e., $(f(k_{i_1}) \geq t_{i_1}) \vee \cdots \vee (f(k_{i_{n_1}}) \geq t_{i_{n_1}}) \vee (f(k_{j_1}) < t_{j_1}) \vee \cdots \vee (f(k_{j_{n_2}}) < t_{j_{n_2}})$, where $\neg$ stands for complement and $n_1 + n_2 = n$. The conjunctive complement construction allows one to search for documents satisfying a condition such as $(f(k_{i_1}) \geq t_{i_1}) \wedge \cdots \wedge (f(k_{i_{n_1}}) \geq t_{i_{n_1}}) \wedge \neg(f(k_{j_1}) \geq t_{j_1}) \wedge \cdots \wedge (f(k_{j_{n_2}}) \geq t_{j_{n_2}})$, i.e., $(f(k_{i_1}) \geq t_{i_1}) \wedge \cdots \wedge (f(k_{i_{n_1}}) \geq t_{i_{n_1}}) \wedge f(k_{j_1}) < t_{j_1}) \wedge \cdots \wedge (f(k_{j_{n_2}}) < t_{j_{n_2}})$.

Furthermore, by combining the above basic constructions, Yi et al. [23] introduced the generic construction for arbitrary threshold query based on keyword frequency.

Like Yi et al.'s solution for the (t, n) threshold query [24], the solutions [23] described here encrypt the thresholds, compare them with the ciphertexts, and store a matching document into the buffer by constructing an encryption of (L, ℓ) linear code of the document. Unlike the (t, n) threshold query solution [24] where only one threshold t is encrypted and enclosed to the searching program, the solutions [23] described here encrypt the frequency threshold for each keyword because different keywords may have different frequency thresholds.

6.2 Overview of Private Searching on Streaming Data

In 2005, Ostrovsky and Skeith [15, 16] gave the first solution for private searching on streaming data as follows.

Assume that the public dictionary of potential keywords is $D = \{w_1, w_2, \cdots, w_{|D|}\}$. To construct a program searching for documents containing one or more of classified keywords $K = \{k_1, k_2, \cdots, k_{|K|}\} \subset D$, the client generates a pair of public and private keys (pk, sk) for a homomorphic encryption scheme $\mathcal{E}$, such as the Paillier cryptosystem [18], and produces an array of ciphertexts $\mathcal{E}(D) = \{c_1, c_2, \cdots, c_{|D|}\}$, one for each keyword $w_i \in D$, such that if $w_i \in K$, then $c_i = \mathcal{E}_{pk}(1)$ and $c_i = \mathcal{E}_{pk}(0)$ otherwise. In addition, the client constructs a buffer $\mathbb{B}$ with γm boxes, each of them is initialized with two ciphertexts $(\mathcal{E}_{pk}(0), \mathcal{E}_{pk}(0))$, where m is the upper bound on the number of matching documents the buffer can accommodate and $m/2^\gamma$ should be negligible.

To perform private searching for keywords, Ostrovsky and Skeith segmented the streaming data S into streaming documents $\{M_1, M_2, \cdots\}$, each of which is composed of a number of words, and filtered one at a time. To process a document M_i, the server, which is provided with $D, \mathcal{E}(D), \mathbb{B}$, computes $d_i = \prod_{w_j \in M_i} c_j = \mathcal{E}_{pk}(|M_i \cap K|)$ and $e_i = d_i^{M_i} = \mathcal{E}_{pk}(M_i \cdot |M_i \cap K|)$, then copies (d_i, e_i) into γ randomly chosen boxes in the buffer $\mathbb{B}$ by multiplying corresponding ciphertexts. If $M_i \cap K = \emptyset$, this step will add an encryption of 0 to each box, having no effect on the corresponding plaintext. If $M_i \cap K \neq \emptyset$, the matching document can be retrieved by computing $M_i = \mathcal{D}_{sk}(e_i)/\mathcal{D}_{sk}(d_i)$ after the buffer returns. If two different matching documents are ever added to the same buffer box, a collision will occur and both copies will be lost. To avoid the loss of matching documents, the buffer size has to be sufficiently large so that each matching document can survive in at least one buffer box with overwhelming probability.

In 2009, Bethencourt et al. [1, 2] proposed a different approach for retrieving matching documents from the buffer. Like the idea of [15], an encrypted dictionary is used, and no-matching documents have no effect on the contents of the buffer. However, rather than using one large buffer and attempting to avoid collisions, Bethencourt, Song, and Water stored the matching documents in three buffers—the data buffer $\mathbb{F}$, the count buffer $\mathbb{C}$, and the matching indices buffer $\mathbb{I}$, and retrieved them by solving linear systems.

Bethencourt et al.'s solution is able to process t documents $\{M_1, M_2, \cdots, M_t\}$ of streaming data. For each document M_i, the server computes d_i and e_i as the Ostrovsky–Skeith protocol and copies d_i and e_i randomly over approximately half of the locations across the buffers $\mathbb{C}$ and $\mathbb{F}$, respectively. A pseudorandom function $g(i, j)$ is used to determine with probability 1/2 whether d_i (or e_i) is copied into a given location j. In addition, the server copies d_i into a fixed number of locations in the buffer $\mathbb{I}$. This is done by using essentially the standard procedure for updating a Bloom filter. Specifically, k hash functions $h_1, h_2, \cdots, h_k$ are used to select the k locations. The locations of $\mathbb{I}$ that d_i is multiplied into are taken to be $h_1(i), h_2(i), \cdots, h_k(i)$.

To retrieve the matching documents, Bethencourt, Song, and Water decrypted three buffers $\mathbb{F}, \mathbb{C}, \mathbb{I}$ to $\mathbb{F}', \mathbb{C}', \mathbb{I}'$ at first. For each of the indices $i \in \{1, 2, ..., t\}$, $h_1(i), h_2(i), \cdots, h_k(i)$ are computed and the corresponding locations in $\mathbb{I}'$ are checked. If all these locations are nonzero, i is added into the list of potential matching indices, denoted as $\{i_1, i_2, \cdots, i_\ell\}$. The values of $c = \{\alpha_{i_1}, \alpha_{i_2}, \cdots, \alpha_{i_\ell}\}$, where $\alpha_{i_j} = |M_{i_j} \cap K|$, are then determined by solving the system of linear equations $A \cdot c^T = \mathbb{C}'^T$, where $A = (g(j, i))$ is an $|\mathbb{C}| \times i_\ell$ matrix. As last step, the content of the matching documents $M' = \{M_{i_1}, M_{i_2}, \cdots, M_{i_\ell}\}$ are determined by solving the system of linear equations $A \cdot diag(c) \cdot M'^T = \mathbb{F}'^T$.

The advantage of Bethencourt et al.'s approach [1, 2], compared to Ostrovsky et al.'s solution [15], is that buffer collisions do not matter because matching documents can be retrieved by solving linear systems. Consequently, the buffer size does not need to be sufficiently large in order to maintain a high probability of recovering all matching documents. In fact, the buffer size becomes optimal, i.e.,

$O(m)$. However, Bethencourt et al.'s approach has a drawback as well. To determine the ordinal numbers of potential matching documents in the decrypted buffer $\mathbb{I}'$, Bethencourt, Song, and Water had to check each of the indices $i \in \{1, 2, \cdots, t\}$ of the data stream. Therefore, the buffer recovering has a running-time proportional to the size of the data stream, i.e., $O(m^{2.376} + t \log(t/m))$. This does not fit the model given by Ostrovsky et al. in [15, 16], in which the buffer is decrypted at the cost which is independent of the stream size.

The idea of private searching for documents containing one or more of keywords can be modified to construct more complicated queries. For example, a query composed of at most a λ AND operations can be performed simply by changing the dictionary D to a dictionary D' containing all $|D|^{\lambda}$ λ-tuples of words in D, which of course comes at a polynomial blow-up of program size.

Using results by Boneh et al. [3], Ostrovsky and Skeith [15, 16] gave a solution for private queries involving an AND of two sets of keywords without increasing the program size. Their basic idea of searching for documents M such that $(M \cap K_1 \neq \emptyset) \wedge (M \cap K_2 \neq \emptyset)$, where K_1, K_2 are two sets of potential keywords, is to construct two arrays of ciphertexts $C_\ell = \{c_1^\ell, c_2^\ell, \cdots, c_{|D|}^\ell\}$ ($\ell = 1, 2$), where c_i^ℓ is the encryption of 1 if $w_i \in K_\ell$ and 0 otherwise. To process a document M, the program computes $v_\ell = \prod_{w_j \cap M} c_j^\ell = \mathcal{E}_{pk}(|M \cap K_\ell|)$ ($\ell = 1, 2$) and then $v = e(v_1, v_2)$, where e is a bilinear map. If $(M \cap K_1 \neq \emptyset) \wedge (M \cap K_2 \neq \emptyset)$ is true, v is an encryption of a nonzero element and 0 otherwise. Then, M is encrypted by replacing 1 with v and 0 with an encryption of 0 and the ciphertext is copied into γ randomly chosen boxes in the buffer $\mathbb{B}$.

Ostrovsky and Skeith [17] showed that the general methods used here to create protocols for searching on streaming data (which are based essentially upon manipulating homomorphic encryption) cannot be extended to perform conjunctive queries beyond what has been accomplished as above. More specifically, if one builds a protocol based on an Abelian group homomorphic encryption, then no conjunctive (of more than one term) can be performed without increasing (super-linearly) the dictionary size. It seems that to make progress in significantly extending the query semantics will likely require fundamentally different approaches to the problem, unless major developments are made in the design of homomorphic encryption scheme.

Gentry [7–10] using lattice-based cryptography constructed the first fully homomorphic encryption scheme. In the same year, Dijk et al. [6] presented a second fully homomorphic encryption scheme. In 2010, Smart et al. [20] presented a refinement of Gentry's scheme giving smaller key and ciphertext sizes. Recent breakthrough in fully homomorphic encryption makes it possible to perform more complicated private queries on streaming data.

In 2012, based on fully homomorphic encryption technique, Yi et al. [24] provided a construction of the searching criteria for private (t, n) threshold query on streaming data, which searches for documents containing more than t out of n keywords, without increasing the dictionary size. Like the idea of [15], an encrypted dictionary $\mathcal{E}(D) = \{c_1, c_2, \cdots, c_{|D|}\}$, where correspondences to n keywords are encryptions of 1 and 0 otherwise, is used. Besides it, an encryption of the threshold t ($\leq |D|$), denoted as $\mathcal{E}_{pk}(t)$, is attached to the program. To process a document M_i, the program computes $d_i = \sum_{w_j \in M_i} c_j = \mathcal{E}_{pk}(|M_i \cap K|)$ and compares $|M_i \cap K|$ with t using d_i and $\mathcal{E}_{pk}(t)$ on the basis of the fully homomorphic property. It outputs a ciphertext α, which is an encryption of 0 if $|M_i \cap K| \geq t$ and an encryption of 1 otherwise. Then M_i is encrypted by replacing 1 with $\alpha + 1$ and 0 with an encryption of 0. The encryption of a matching document is stored into the buffer by constructing an encryption of (L, ℓ) linear code of the document, where ℓ and L are the plain document size and the plain buffer size, respectively, and then position-wise adding the code into the buffer. To keep up to m matching documents, the buffer size only needs to be $m\ell k$ ($= Lk$), where k is a security parameter. In addition, the computational decoding cost is $O(m\ell k^2)$ independent of the streaming size. Furthermore, the buffer can keep at most m matching documents. In case there are more than m matching documents in the streaming data, the buffer stores the first m matching documents and throws the rest away. Thus, the buffer collision is no longer an issue.

6.3 Preliminaries

6.3.1 Integer Addition with FHE

In general, a fully homomorphic encryption scheme $\mathcal{E}$ has the following properties:

$$\mathcal{E}(m_1) + \mathcal{E}(m_2) = \mathcal{E}(m_1 \oplus m_2),$$

$$\mathcal{E}(m_1)\mathcal{E}(m_2) = \mathcal{E}(m_1 m_2),$$

for any $m_1, m_2 \in \{0, 1\}$.

Based on the above two properties, given $\mathcal{E}(m_1)$ and $\mathcal{E}(m_2)$, we can construct

$$\mathcal{E}(m_1 \wedge m_2) = \mathcal{E}(m_1)\mathcal{E}(m_2),$$

$$\mathcal{E}(m_1 \vee m_2) = \mathcal{E}(m_1) + \mathcal{E}(m_2) + \mathcal{E}(m_1)\mathcal{E}(m_2),$$

for any $m_1, m_2 \in \{0, 1\}$.

For a positive integer $M = (m_1 m_2 \cdots m_\ell)_b$ (a binary expression), we write $\mathcal{E}(M) = (\mathcal{E}(m_1), \mathcal{E}(m_2), \cdots, \mathcal{E}(m_\ell))$. Given $\mathcal{E}(M_1) = (\mathcal{E}(x_1), \mathcal{E}(x_2), \cdots, \mathcal{E}(x_\ell))$ and $\mathcal{E}(M_2) = (\mathcal{E}(y_1), \mathcal{E}(y_2), \cdots, \mathcal{E}(y_\ell))$, we can construct $\mathcal{E}(M_1 + M_2)$ as follows:

Assume that $(x_1 x_2 \cdots x_\ell)_b + (y_1 y_2 \cdots y_\ell)_b = (z_0 z_1 \cdots z_\ell)_b$ where z_0 is the carry bit. On the basis of the digital circuit for binary integer addition [19], we have

$$c_{i-1} = x_i y_i \vee (x_i \oplus y_i) c_i$$

$$z_i = x_i \oplus y_i \oplus c_i$$

for $i = \ell, \cdots, 2, 1$, where $c_\ell = 0$ and $z_0 = c_0$. Due to $\alpha \vee \beta = (\alpha \oplus \beta) \oplus (\alpha\beta)$, one can compute

$$\mathcal{E}(a_{i-1}) = \mathcal{E}(x_i)\mathcal{E}(y_i) = \mathcal{E}(x_i \oplus y_i)$$
$$\mathcal{E}(b_{i-1}) = (\mathcal{E}(x_i) + \mathcal{E}(y_i))\mathcal{E}(c_i) = \mathcal{E}((x_i \oplus y_i)c_i)$$
$$\mathcal{E}(c_{i-1}) = (\mathcal{E}(a_{i-1}) + \mathcal{E}(b_{i-1})) + \mathcal{E}(a_{i-1})\mathcal{E}(b_{i-1})$$
$$= \mathcal{E}((a_{i-1} \oplus b_{i-1}) \oplus a_{i-1}b_{i-1})$$
$$\mathcal{E}(z_i) = \mathcal{E}(x_i) + \mathcal{E}(y_i) + \mathcal{E}(c_i) = \mathcal{E}(x_i \oplus y_i \oplus c_i)$$

for $i = \ell, \cdots, 2, 1$, then let $\mathcal{E}(z_0) = \mathcal{E}(c_0)$ and $\mathcal{E}(M_1 + M_2) = (\mathcal{E}(z_0), \mathcal{E}(z_1), \cdots, \mathcal{E}(z_\ell))$. We define $\mathcal{E}(M_1) \boxplus \mathcal{E}(M_2) = \mathcal{E}(M_1 + M_2)$.

6.3.2 Integer Comparison with FHE

In particular, given $\mathcal{E}(M_1)$ and $\mathcal{E}(M_2)$ where M_1 and M_2 are two positive integers, we can compare M_1 with M_2 by computing

$$\mathcal{E}(\overline{M}_1) \boxplus \mathcal{E}(\overline{-M}_2) = \mathcal{E}(\overline{M}_1 + \overline{-M}_2)$$

where $\overline{M}_1$ and $\overline{-M}_2$ are two's complements of M_1 and $-M_2$, respectively. Two's complement system is the most common method of representing signed integers on computers (please refer to [12, 13, 22]).

If $M_1 \geq M_2$, the most significant bit of $\overline{M}_1 + \overline{-M}_2$ is 0 and 1 otherwise.

Given $\mathcal{E}(M) = (\mathcal{E}(m_1), \mathcal{E}(m_2), \cdots, \mathcal{E}(m_\ell))$, we have

$$\mathcal{E}(\overline{M}) = (\mathcal{E}(0), \mathcal{E}(m_1), \mathcal{E}(m_2), \cdots, \mathcal{E}(m_\ell)),$$
$$\mathcal{E}(\overline{-M}) = (\mathcal{E}(1), \mathcal{E}(m_1) + 1, \mathcal{E}(m_2) + 1, \cdots, \mathcal{E}(m_\ell) + 1) \boxplus \mathcal{E}(1).$$

6.3.3 Binary Linear Codes

An $[n, k]$ binary linear code C of length n and dimension k is a k-dimensional subspace of F_2^n according to [14]. A generator matrix for C is a $k \times n$ matrix

$$G = \begin{pmatrix} a_{11} & a_{12} & \cdots & a_{1n} \\ a_{21} & a_{22} & \cdots & a_{2n} \\ \cdots & \cdots & \cdots & \cdots \\ a_{k1} & a_{k2} & \cdots & a_{kn} \end{pmatrix}$$

where $a_{ij} \in F_2$, such that $C = \{(b_1, b_2, \cdots, b_k)G \mid b_i \in F_2\}$. The matrix G corresponds to a map $F_2^k \to F_2^n$ expanding a message $(b_1, b_2, \cdots, b_k)$ of length k to an n-bit string.

We say that binary linear codes $C_1, C_2, \cdots, C_m$ are orthogonal if $C_i \cap C_j = \emptyset$ and $c_i \cdot c_j = 0$ for any two codewords $c_i \in C_i$ and $c_j \in C_j$ ($i, j = 1, 2, \cdots, m, i \neq j$), where "$\cdot$" stands for the dot product operation. In case where $m = n/k$, there exist m simple orthogonal binary linear codes $C_1, C_2, \cdots, C_m$. The generator matrix of C_i is

$$G_i = \begin{pmatrix} \cdots 1\ 0 \cdots 0 \cdots 0\ 0 \cdots 0 \\ \cdots 0\ 1 \cdots 0 \cdots 0\ 0 \cdots 0 \\ \cdots \cdots \cdots \cdots \cdots \cdots \cdots \cdots \\ \cdots 0\ 0 \cdots 1 \cdots 0\ 0 \cdots 0 \end{pmatrix}$$

where the element at $(j, (i-1)k + j)$ (for $i = 1, 2, \cdots, m$ and $j = 1, 2, \cdots, k$) is 1 and otherwise 0.

6.4   Definitions

Definitions for general private queries were given in [15, 16]. In this chapter, slightly different definitions are given based on the paper by Yi et al. [23].

Like the streaming model given in [15, 16], we consider a universe of words $W = \{0, 1\}^*$ and a dictionary $D \subset W$ with $|D| < \infty$. We think of a document M just to be an ordered, finite sequence of words in W and a stream of documents S just to be any sequence of documents. We define a set of keywords to be any subset $K \subset D$.

Definition 6.1. A query Q over a set of keywords K, denoted as Q_K, is a logical expression of keywords in K.

Definition 6.2. Given a document M and a query Q_K, we define $Q_K(M) = 1$ if M matches the query Q_K and $Q_K(M) = 0$ otherwise.

Definition 6.3. For a query Q_K, a private query protocol is composed of the following probabilistic polynomial time algorithms:

1. KeyGen(k): Takes a security parameter k and generates a pair of public and secret keys (pk, sk).
2. FilterGen(D, Q_K, pk): Takes a dictionary D, a query Q_K, the public key pk, and generates a program F.
3. FilterExec(S, F, pk, m): Takes a stream of documents S, F searches for any document $M \in S$ such that $Q_K(M) = 1$ (processing one document at a time), encrypts each matching document with the public key pk, keeps up to m encrypted matching document in a buffer $\mathbb{B}$, and finally outputs an encrypted buffer $\mathbb{B}$.

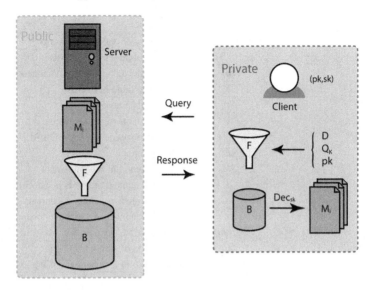

Fig. 6.1 Model for private searching on streaming data

4. BufferDec($\mathbb{B}$, sk): Decrypts the encrypted buffer $\mathbb{B}$, produced by F as above, using the private key sk and outputs a plain buffer $\mathbb{B}^*$, a collection of the matching documents from S.

Based on Definition 6.3, the model for privacy searching on stream data can be illustrated in Fig. 6.1.

Definition 6.4 (Correctness of Private Query Protocol). Let $F = \mathsf{FilterExec}(S, F, pk, m)$, where D is a dictionary, $\mathcal{Q}_K$ is a query over keywords K, $(pk, sk) = \mathsf{KeyGen}(k)$, and m is an upper bound on the number of matching documents; we say that a private query protocol is correct if the following holds: Let F run on any document stream S, $\mathbb{B} = F(S)$, $\mathbb{B}^* = \mathsf{BufferDec}(\mathbb{B}, sk)$.

1. (Compiled Program Conciseness) $|F| = O(|D|)$
2. (Output Conciseness) $|\mathbb{B}| = O(m)$
3. (Search Completeness) If $|\{M \in S | \mathcal{Q}_K(M) = 1\}| \leq m$, then
   $\mathbb{B}^* = \{M \in S | \mathcal{Q}_K(M) = 1\}$.
4. (Collision Freeness) If $|\{M \in S | \mathcal{Q}_K(M) = 1\}| > m$, then
   $|\mathbb{B}^* \cap \{M \in S | \mathcal{Q}_K(M) = 1\}| = m$.

where the probabilities are taken over all coin-tosses of F, $\mathsf{FilterGen}$, and KeyGen.

Definition 6.5 (Privacy). Fix a dictionary D. Consider the following game between an adversary $\mathcal{A}$, and a challenger $\mathcal{C}$. The game consists of the following steps:

Table 6.1 Notations

Symbol	Explanation
D	Dictionary of possible keywords
$\lvert D \rvert$	Number of possible keywords in D
w_i	Word in the dictionary and documents
K	Set of classified keywords
k_i	Classified keyword
n	Number of classified keywords
Q_K	Logical expression of keywords in K
F	Filter program
M, M_i	Document in the streaming data
d	Maximal number of words in a document
$\mathbb{B}$	Buffer to store matching documents
m	Maximal number of matching documents in $\mathbb{B}$
(pk, sk)	Public/private key pair
$\mathcal{E}_{pk}(b)$	Encryption of a bit b using pk
$\mathcal{D}_{sk}(c)$	Decryption of a ciphertext c using sk
$\widehat{0}, \widehat{1}$	Encryptions of 0 and 1 using pk
$\lvert C \rvert$	Size of the ciphertext
$f(k_i)$	Frequency of keyword k_i in a document
t_i	Frequency threshold of keyword k_i
$\widehat{w_i}$	Encryption of frequency threshold t_i
$\overline{t}$	Two's complement of an integer t
$\boxplus$	Homomorphic addition of integers
$\neg()$	Complement of a condition

1. The challenger C first runs KeyGen(k) to obtain a pair of public and secret keys (pk, sk) and then sends pk and m, the upper bound on the number of matching documents, to $\mathcal{A}$.
2. The adversary $\mathcal{A}$ chooses two queries for two sets of keywords, Q_{0K_0}, Q_{1K_1}, with $K_0, K_1 \subset D$ and sends them to C.
3. The challenger C chooses a random bit $b \in \{0, 1\}$ and executes FilterGen(D, Q_{bK_b}, pk) to create F_b, the filtering program for the query Q_{bK_b}, and then sends F_b back to $\mathcal{A}$.
4. The adversary $\mathcal{A}(F_b, pk, m)$ can experiment with code of F_b in an arbitrary non-black-box way and finally output $b' \in \{0, 1\}$.

The adversary wins the game if $b' = b$ and loses otherwise. We define the adversary $\mathcal{A}$'s advantage in this game to be $\mathsf{Adv}_{\mathcal{A}}(k) = \lvert \Pr(b' = b) - 1/2 \rvert$. We say that a private query protocol is semantically secure if for any probabilistic polynomial time (PPT) adversary $\mathcal{A}$, we have that $\mathsf{Adv}_{\mathcal{A}}(k)$ is a negligible function, where the probability is taken over coin-tosses of the challenger and the adversary.

In the rest of this chapter, we will use the notations as listed in Table 6.1.

6.5 Private Threshold Query Based on Keyword Frequency

6.5.1 Disjunctive Threshold Query

Formally, a disjunctive threshold query over keywords $K = \{k_1, k_2, \cdots, k_n\}$ can be expressed as

$$\mathcal{Q}_K = (f(k_1) \geq t_1) \vee (f(k_2) \geq t_2) \vee \cdots \vee (f(k_n) \geq t_n)$$

where $f(k_i)$ $(1 \leq i \leq n)$ is the frequency of the keyword k_i in the document and t_i is the given threshold. It is easy to see

Lemma 6.6 ([23]). *Given a document M, a disjunctive threshold query $\mathcal{Q}_K(M) = 1$ if and only if there exists i such that $f(k_i) \geq t_i$.*

Following the model described in Sect. 6.4, the protocol for disjunctive threshold queries is composed of four algorithms KeyGen, FilterGen, FilterExec, *and* BufferDec. The construction is based on a fully homomorphic encryption scheme and can be formally presented as follows.

Key Generation

KeyGen(k): Run the key generation algorithm for the underlying fully homomorphic encryption scheme to produce the private key sk and the public key pk.

Filter Program Generation

FilterGen($D, \mathcal{Q}_K, pk$): This algorithm outputs a filter program F for disjunctive threshold query $\mathcal{Q}_K$ based on keyword frequency.

Assume that the public dictionary $D = \{w_1, w_2, \cdots, w_{|D|}\}$, keywords $K = \{k_1, k_2, \cdots, k_n\} \subset D$, $d = \lceil \log_2 |M| \rceil$ where $|M|$ stands for the maximal number of words the document M may contain, then F consists of the dictionary D, disjunctive query sign (denoted as 00), and an array of ciphertexts

$$\hat{D} = \{\hat{w}_1, \hat{w}_2, \cdots, \hat{w}_{|D|}\},$$

where $\hat{w}_i = \mathcal{E}_{pk}(t_i)$ and

$$t_i = \begin{cases} frequency\ threshold\ for\ k_j & \text{if } w_i = k_j \in K \\ 2^d - 1 & \text{if } w_i \notin K \end{cases}$$

Remark. Because the document M contains at most $2^d - 1$ words, the frequency of any word in M is less than $2^d - 1$. In practice, a document which repeats a word for $2^d - 1$ times is unusual. We do not consider this special case in this chapter. We set the frequent threshold of a non-keyword as $2^d - 1$ so that its frequency in M is never more than the threshold.

Assume $t_i = (a_{i1}a_{i2}\cdots a_{id})_b$ where $a_{ij} \in \{0,1\}$, then $\hat{w}_i = \mathcal{E}_{pk}(t_i) = (\mathcal{E}_{pk}(a_{i1}), \mathcal{E}_{pk}(a_{i2}), \cdots, \mathcal{E}_{pk}(a_{id}))$. The array of ciphertexts $\hat{D}$ contains n encryptions of frequency thresholds and $|D| - n$ encryptions of $2^d - 1$.

Filter Program Execution

FilterExec(S, F, pk, m): This algorithm outputs an encrypted buffer $\mathbb{B}$ keeping up to m matching documents.

First of all, the program F constructs a data buffer $\mathbb{B}$ with $m\ell$ boxes, each of them is initialized with $\mathcal{E}_{pk}(0)$, where ℓ is the size of the document. Next, F constructs a base buffer $\mathbb{G}$ with m boxes, which are initialized with $(\mathcal{E}_{pk}(0), \cdots, \mathcal{E}_{pk}(0), \mathcal{E}_{pk}(1))$.

Remark. The data buffer $\mathbb{B}$ is used to store the matching documents and the base buffer $\mathbb{G}$ is used to ensure the first m matching documents are stored in $\mathbb{B}$ without collision.

In addition, the program F constructs the encryption of the two's complement of $-t_i$ (denoted as $\overline{-t_i}$) with $\hat{w}_i = \mathcal{E}_{pk}(t_i)$, that is,

$$\mathcal{E}_{pk}(\overline{-t_i}) = (\mathcal{E}_{pk}(1), \mathcal{E}_{pk}(a_{i1}) + 1, \cdots, \mathcal{E}_{pk}(a_{id}) + 1) \boxplus \mathcal{E}_{pk}(1)$$

The leftmost bit of the two's complement of a negative integer is 1 and 0 otherwise.

Upon receiving an input document $M = (m_1 m_2 \cdots m_\ell)_b$ from the stream S, in order to determine if M is a matching document or not, the program F homomorphically compute a ciphertext $\mathcal{E}_{pk}(c_0)$ such that M is a matching document if $c_0 = 1$ and 0 otherwise. It proceeds with the following steps:

1. (*Word Collection*) The program F first collects

$$\hat{H} = \{w_i, f(w_i)|w_i \in M \cap D\}$$

where $f(w_i)$ is the frequency of w_i in the document M.

Remark. $\hat{H}$ is the set of common words in the document M and the dictionary D and their frequencies in M.

Next, F constructs the encryption of the two's complement of $f(w_i) = (b_{i1}b_{i2}\cdots b_{id})_b$, denoted as $\overline{f(w_i)}$, for each $w_i \in \hat{H}$, that is,

$$\mathcal{E}_{pk}(\overline{f(w_i)}) = (\mathcal{E}_{pk}(0), \mathcal{E}_{pk}(b_{i1}), \mathcal{E}_{pk}(b_{i2}), \cdots, \mathcal{E}_{pk}(b_{id}))$$

Remark. Because $f(w_i) < 2^d - 1$, we only consider the encryptions of the d bits and one sign bit.

2. (*Frequency Comparison*) For each $w_i \in \hat{H}$, the program F homomorphically compares the frequency $f(w_i)$ and the frequency threshold t_i by computing

$$\mathcal{E}_{pk}(\overline{f(w_i)} + \overline{-t_i})$$
$$= \mathcal{E}_{pk}(\overline{f(w_i)}) \boxplus \mathcal{E}_{pk}(\overline{-t_i})$$
$$= (\mathcal{E}_{pk}(c_{i0}), \mathcal{E}_{pk}(c_{i1}), \mathcal{E}_{pk}(c_{i2}), \cdots, \mathcal{E}_{pk}(c_{id}))$$

from which only $\mathcal{E}_{pk}(c_{i0})$ is extracted. In two's complement system, if $c_{i0} = 0$, then $f(w_i) \geq t_i$ and otherwise $f(w_i) < t_i$.

Next, the program F computes

$$\mathcal{E}_{pk}(c_0) = \mathcal{E}_{pk}\Big(\bigvee_{w_i \in \hat{H}} (c_{i0} \oplus 1)\Big) \tag{6.1}$$

by repeatedly using $\mathcal{E}_{pk}(c_{i0} \vee s) = \mathcal{E}_{pk}(c_{i0}) + \mathcal{E}_{pk}(s) + \mathcal{E}_{pk}(c_{i0})\mathcal{E}_{pk}(s)$.

If $c_0 = 1$, then there exists i such that $c_{i0} \oplus 1 = 1$ (i.e., $c_{i0} = 0$ and $f(w_i) \geq t_i$). If $w_i \notin K$, then $t_i = 2^d - 1$ and it is impossible that $f(w_i) \geq 2^d - 1$. This means that $w_i \in K$ and $f(w_i) \geq t_i$. According to Lemma 6.6, M is a matching document.

If $c_0 = 0$, then $c_{i0} \oplus 1 = 0$ (i.e., $c_{i0} = 1$ and $f(w_i) < t_i$) for all $w_i \in M \cap D$. According to Lemma 6.6, M is not a matching document.

3. (*Document Storing*) Assume that the state of the base buffer $\mathbb{G}$ is $(\hat{g}_m, \hat{g}_{m-1}, \cdots, \hat{g}_1)$, where $\hat{g}_i$ is an encryption of either 0 or 1, the program F constructs an encrypted $\ell \times L$ generator matrix G for an $[L, \ell]$ binary linear code as follows:

$$G = \begin{pmatrix} \hat{g}_1 & \hat{0} & \cdots & \hat{0} & \cdots & \hat{g}_m & \hat{0} & \cdots & \hat{0} \\ \hat{0} & \hat{g}_1 & \cdots & \hat{0} & \cdots & \hat{0} & \hat{g}_m & \cdots & \hat{0} \\ \cdot & \cdot & \cdots & \cdot & \cdots & \cdot & \cdot & \cdots & \cdot \\ \hat{0} & \hat{0} & \cdots & \hat{g}_1 & \cdots & \hat{0} & \hat{0} & \cdots & \hat{g}_m \end{pmatrix}$$

where $L = m\ell$ and the element at $(i, (j-1)\ell + i)$ (for $i = 1, 2, \cdots, \ell$ and $j = 1, 2, \cdots, m$) is $\hat{g}_j$ and otherwise $\hat{0}$ (an encryption of 0).

To store the encryption of the document M into the data buffer $\mathbb{B}$, the program F computes

$$\hat{M} = \mathcal{E}_{pk}(c_0)\mathcal{E}_{pk}(M)G$$
$$= (\mathcal{E}_{pk}(c_0 m_1), \cdots, \mathcal{E}_{pk}(c_0 m_\ell))G$$

and position-wise adds the result into the data buffer $\mathbb{B}$, denoted as

$$\mathbb{B} = \mathbb{B} + \hat{M}$$

If $c_0 = 1$, then $\hat{M}$, the encryption of the binary linear code of the matching document M, is kept in the data buffer $\mathbb{B}$. If $c_0 = 0$, then $\hat{M}$ is the encryption of 0, which has no effect on the data buffer $\mathbb{B}$.

4. In order to avoid collision when storing next matching document into the data
   buffer $\mathbb{B}$, the program F updates the base buffer $\mathbb{G}$ by homomorphically shifting
   $\mathcal{E}_{pk}(1)$ in the base buffer $\mathbb{G}$ by one position to the left if M is a matching
   document and 0 position otherwise. This is done by computing

$$\mathbb{G}' = \mathbb{G} \boxplus \mathcal{E}_{pk}(c_0)\mathbb{G}$$

where $\mathbb{G}$ is treated as the encryption of an m-bit integer and replacing $\mathbb{G}$ with $\mathbb{G}'$.

Remark. Initially, $\mathbb{G} = (\mathcal{E}_{pk}(0), \cdots, \mathcal{E}_{pk}(0), \mathcal{E}_{pk}(1))$. If $c_0 = 0$, the buffer does
not change. Only when $c_0 = 1$, the buffer is updated by shifting $\mathcal{E}_{pk}(1)$ one
position to the left. We only consider the encryptions of the right m bits. After
shifting m times, the buffer becomes the encryptions of all zeros. The buffer
contains at most one encryption of 1 all the time.

Buffer Decryption
BufferDec($\mathbb{B}, sk$): Using the secret key sk, the algorithm decrypts the encrypted
data buffer $\mathbb{B}$, sent back by the filter program F, one box at a time. Assume that
the decrypted data buffer is $(m'_1 m'_2 \cdots m'_L)_b$ where $L = m\ell$, then the set of matching
documents is

$$\mathbb{B}^* = \{M = (m'_{i\ell+1} \cdots m'_{i\ell+\ell})_b | M \neq 0, i = 0, 1, \cdots, m - 1\}$$

Correctness: The filter program F is composed of D (the dictionary) and $\hat{D}$
(the encryption of the frequency thresholds). The size of $\hat{D}$ is $|D|dk$, where k is
the security parameter. Therefore, the size of the filter program $|F| = O(|D|)$.

The data buffer $\mathbb{B}$ has $m\ell$ boxes (where ℓ is the size of the document), each keeps
a ciphertext of one bit. The size of the buffer $|\mathbb{B}| = m\ell k = O(m)$.

We need to show that if the number of matching documents is less than or equal
to m, then $\mathbb{B}^* = \{M \in S | \mathcal{Q}_K(M) = 1\}$ (search completeness) and otherwise we
have $|\mathbb{B}^* \cap \{M \in S | \mathcal{Q}_K(M) = 1\}| = m$ (collision freeness).

Assume that the matching documents in the stream $S = \{M_1, M_2, \cdots, \}$ are
$\{M_{i_1}, M_{i_2}, \cdots\}$. Initially, the data buffer $\mathbb{B} = (\mathcal{E}_{pk}(0), \mathcal{E}_{pk}(0), \cdots, \mathcal{E}_{pk}(0))$, the base
buffer $\mathbb{G} = (\mathcal{E}_{pk}(0), \cdots, \mathcal{E}_{pk}(0), \mathcal{E}_{pk}(1))$, and the generator matrix

$$G = \begin{pmatrix} \hat{1} & \hat{0} & \cdots & \hat{0} & \cdots & \hat{0} & \hat{0} & \cdots & \hat{0} \\ \hat{0} & \hat{1} & \cdots & \hat{0} & \cdots & \hat{0} & \hat{0} & \cdots & \hat{0} \\ \cdot & \cdot & \cdots & \cdot & \cdot & \cdot & \cdot & \cdots & \cdot \\ \hat{0} & \hat{0} & \cdots & \hat{1} & \cdots & \hat{0} & \hat{0} & \cdots & \hat{0} \end{pmatrix}$$

where $\hat{1}$ and $\hat{0}$ are encryptions of 1 and 0, respectively.

For a non-matching document M, we have $c_0 = 0$ and thus $\hat{M} = \mathcal{E}_{pk}(c_0)\mathcal{E}_{pk}(M)G = (\mathcal{E}_{pk}(0), \mathcal{E}_{pk}(0), \cdots, , \mathcal{E}_{pk}(0))$, the data buffer $\mathbb{B} = \mathbb{B} + \hat{M} = \mathbb{B}$ and the base buffer $\mathbb{G}' = \mathbb{G} \boxplus \mathcal{E}_{pk}(c_0)\mathbb{G} = \mathbb{G}$, which means that the content of $\mathbb{B}$
and $\mathbb{G}$ do not change.

When the filter program F deals with the matching document M_{i_j} ($1 \leq j \leq m$), we have $c_0 = 1$ and the state of the base buffer $\mathbb{G}$ is evolved from $(\mathcal{E}_{pk}(0), \cdots, \mathcal{E}_{pk}(0), \mathcal{E}_{pk}(1))$ by shifting $\mathcal{E}_{pk}(1)$ to the left $j - 1$ positions because there are $j - 1$ matching documents before M_{i_j}. Therefore, the generator matrix

$$G = \begin{pmatrix} \cdots \hat{1} \, \hat{0} \cdots \hat{0} \cdots \hat{0} \, \hat{0} \cdots \hat{0} \\ \cdots \hat{0} \, \hat{1} \cdots \hat{0} \cdots \hat{0} \, \hat{0} \cdots \hat{0} \\ \cdots \, \cdot \, \cdot \, \cdots \, \cdot \, \cdots \, \cdot \, \cdot \, \cdots \, \cdot \\ \cdots \hat{0} \, \hat{0} \cdots \hat{1} \cdots \hat{0} \, \hat{0} \cdots \hat{0} \end{pmatrix}$$

and $\hat{M}_{i_j} = \mathcal{E}_{pk}(c_0)\mathcal{E}_{pk}(M_{i_j})G = (\mathcal{E}_{pk}(0), \cdots, \mathcal{E}_{pk}(M_{i_j}), \cdots, \mathcal{E}_{pk}(0))$ and $\mathbb{B} = \mathbb{B} + \hat{M}_{i_j} = (\mathcal{E}_{pk}(M_{i_1}), \cdots, \mathcal{E}_{pk}(M_{i_{j-1}}), \mathcal{E}_{pk}(M_{i_j}), \mathcal{E}_{pk}(0), \cdots, \mathcal{E}_{pk}(0))$. After that, the base buffer $\mathbb{G}$ is updated to $\mathbb{G} \boxplus \mathcal{E}_{pk}(c_0)\mathbb{G} = \mathbb{G} \boxplus \mathbb{G}$, i.e., shifting $\mathcal{E}_{pk}(1)$ further to the left by one position.

In case when the filter program F deals with the matching document M_{i_j} ($j > m$), although $c_0 = 1$, the base buffer $\mathbb{G}$ contains the encryptions of all zeros and so does the generator matrix G. Therefore, $\hat{M}_{i_j} = \mathcal{E}_{pk}(c_0)\mathcal{E}_{pk}(M_{i_j})G = (\mathcal{E}_{pk}(0), \mathcal{E}_{pk}(0), \cdots, \mathcal{E}_{pk}(0))$ and $\mathbb{B} = \mathbb{B} + \hat{M}_{i_j} = \mathbb{B}$. This means the matching document M_{i_j} ($j > m$) has no effect on the data buffer $\mathbb{B}$.

In summary, both search completeness and collision freeness are true.

6.5.2   Conjunctive Threshold Query

Formally, a conjunctive threshold query over keywords $K = \{k_1, k_2, \cdots, k_n\}$ can be expressed as

$$\mathcal{Q}_K = (f(k_1) \geq t_1) \wedge (f(k_2) \geq t_2) \wedge \cdots \wedge (f(k_n) \geq t_n)$$

where $f(k_i)$ ($1 \leq i \leq n$) is the frequency of the keyword k_i in the document and t_i is the given threshold. It is easy to see

Lemma 6.7 ([23]). *Given a document M, a conjunctive threshold query $\mathcal{Q}_K(M) = 1$ if and only if $f(k_i) \geq t_i$ for $1 \leq i \leq n$.*

Following the model described in Sect. 6.4, the protocol of conjunctive threshold query is composed of four algorithms KeyGen, FilterGen, FilterExec, and BufferDec. The conjunctive construction can be formally presented as follows.

Key Generation
KeyGen(k): Run the key generation algorithm for the underlying fully homomorphic encryption scheme to produce the private key sk and the public key pk.

Filter Program Generation

FilterGen($D, \mathcal{Q}_K, pk$): This algorithm outputs a filter program F for conjunctive threshold query $\mathcal{Q}_K$ based on keyword frequency.

Assume that the public dictionary $D = \{w_1, w_2, \cdots, w_{|D|}\}$, keywords $K = \{k_1, k_2, \cdots, k_n\} \subset D$, $d = \lceil \log_2 |M| \rceil$ where $|M|$ stands for the maximal number of words the document M can contain, then F consists of the dictionary D, conjunctive query sign (denoted as 01), and an array of ciphertexts

$$\hat{D} = \{\hat{w}_1, \hat{w}_2, \cdots, \hat{w}_{|D|}\},$$

where $\hat{w}_i = \mathcal{E}_{pk}(t_i)$ and

$$t_i = \begin{cases} \text{frequency threshold for } k_j & \text{if } w_i = k_j \in K \\ 0 & \text{if } w_i \notin K \end{cases}$$

Remark. Because the document M contains at most $2^d - 1$ words, both t_i and $t = \sum_{w_i \in K} t_i$ must be less than $2^d - 1$. We set the frequent threshold of a non-keyword as 0 so that its frequency is always more than the threshold.

Assume $t_i = (a_{i1}a_{i2}\cdots a_{id})_b$ where $a_{ij} \in \{0, 1\}$, then $\hat{w}_i = \mathcal{E}_{pk}(t_i) = (\mathcal{E}_{pk}(a_{i1}), \mathcal{E}_{pk}(a_{i2}), \cdots, \mathcal{E}_{pk}(a_{id}))$. The array of ciphertexts contains n encryptions of frequency thresholds and $|D| - n$ encryptions of 0.

Filter Program Execution

FilterExec(S, F, pk, m): This algorithm outputs an encrypted buffer $\mathbb{B}$ keeping up to m matching documents.

First of all, the program F constructs a data buffer $\mathbb{B}$ with $m\ell$ boxes, each of them is initialized with $\mathcal{E}_{pk}(0)$. Next, F constructs a base buffer $\mathbb{G}$ with m boxes, which are initialized with $(\mathcal{E}_{pk}(0), \cdots, \mathcal{E}_{pk}(0), \mathcal{E}_{pk}(1))$. In addition, the program F constructs the encryption of the two's complement of $-t_i$ (denoted as $\overline{-t_i}$) with $\hat{w}_i = \mathcal{E}_{pk}(t_i)$, that is,

$$\mathcal{E}_{pk}(\overline{-t_i}) = (\mathcal{E}_{pk}(1), \mathcal{E}_{pk}(a_{i1}), \cdots, \mathcal{E}_{pk}(a_{id})) \boxplus \mathcal{E}_{pk}(1),$$

and the encryption of $t = \sum_{w_i \in K} t_i$ with $\hat{D}$ (please note that $t_i = 0$ when $w_i \notin K$), that is,

$$\boxplus_{i=1}^{|D|} \hat{w}_i = \hat{w}_1 \boxplus \hat{w}_2 \boxplus \cdots \boxplus \hat{w}_{|D|},$$

and the encryption of the two's complement of $-t$ (denoted as $\overline{-t}$) with $\mathcal{E}_{pk}(t)$, that is,

$$\mathcal{E}_{pk}(\overline{-t}) = (\mathcal{E}_{pk}(1), \mathcal{E}_{pk}(\alpha_1), \cdots, \mathcal{E}_{pk}(\alpha_d)) \boxplus \mathcal{E}_{pk}(1).$$

Upon receiving an input document $M = (m_1 m_2 \cdots m_\ell)_b$ from the stream S, in order to determine if M is a matching document or not, the program F homomorphically computes a ciphertext $\mathcal{E}_{pk}(c_0)$ such that M is a matching document if $c_0 = 1$ and 0 otherwise. It proceeds with the following steps:

1. (*Word Collection*) The program F first collects

$$\hat{H} = \{w_i, f(w_i) | w_i \in M \cap D\}$$

where $f(w_i)$ is the frequency of w_i in the document M. Next, F constructs the encryption of the two's complement of $f(w_i) = (b_{i1}b_{i2} \cdots b_{id})_b$, denoted as $\overline{f(w_i)}$, for each $w_i \in \hat{H}$, that is,

$$\mathcal{E}_{pk}(\overline{f(w_i)}) = (\mathcal{E}_{pk}(0), \mathcal{E}_{pk}(b_{i1}), \mathcal{E}_{pk}(b_{i2}), \cdots, \mathcal{E}_{pk}(b_{id})),$$

and the encryption of the two's complement of $t' = \sum_{w_i \in \hat{H}} t_i = (\beta_1, \beta_2, \cdots, \beta_d)$, denoted as $\overline{t'}$, that is,

$$\mathcal{E}_{pk}(\overline{t'}) = (\mathcal{E}_{pk}(0), \mathcal{E}_{pk}(\beta_1), \mathcal{E}_{pk}(\beta_2), \cdots, \mathcal{E}_{pk}(\beta_d)),$$

Remark. $\mathcal{E}_{pk}(t') = (\mathcal{E}_{pk}(\beta_1), \mathcal{E}_{pk}(\beta_2), \cdots, \mathcal{E}_{pk}(\beta_d))$ can be obtained with $\boxplus_{w_i \in \hat{H}} \hat{w}_i$. Because the sum t' is never more than $2^d - 1$, we consider d bits of t' only.

2. (*Frequency Comparison*) For each $w_i \in \hat{H}$, the program F homomorphically compares $f(w_i)$ and t_i by computing

$$\mathcal{E}_{pk}(\overline{f(w_i)} + \overline{-t_i})$$
$$= \mathcal{E}_{pk}(\overline{f(w_i)}) \boxplus \mathcal{E}_{pk}(\overline{-t_i})$$
$$= (\mathcal{E}_{pk}(c_{i0}), \mathcal{E}_{pk}(c_{i1}), \mathcal{E}_{pk}(c_{i2}), \cdots, \mathcal{E}_{pk}(c_{id}))$$

from which only $\mathcal{E}_{pk}(c_{i0})$ is extracted. If $c_{i0} = 0$, then $f(w_i) \geq t_i$ and otherwise $f(w_i) < t_i$.

In addition, the program F homomorphically checks if the document M contains all keywords in K by computing

$$\mathcal{E}_{pk}(\overline{t'} + \overline{-t})$$
$$= \mathcal{E}_{pk}(\overline{t'}) \boxplus \mathcal{E}_{pk}(\overline{-t})$$
$$= (\mathcal{E}_{pk}(\gamma_0), \mathcal{E}_{pk}(\gamma_1), \mathcal{E}_{pk}(\gamma_2), \cdots, \mathcal{E}_{pk}(\gamma_d))$$

from which only $\mathcal{E}_{pk}(\gamma_0)$ is extracted. If $\gamma_0 = 0$, then $t' \geq t$ and thus $t' = t$ and the document contains all keywords in K. If $\gamma_0 = 1$, then $t' < t$ and the document does not contain all keywords in K.

Remark. Because $t' = \sum_{w_i \in \hat{H}} t_i = \sum_{w_i \in \hat{H} \cap K} t_i \leq \sum_{w_i \in K} t_i = t$, the inequality $t' \geq t$ means that $t' = t$, $\hat{H} \cap K = K$, and the document contains all keywords in K. Reversely, the inequality $t' < t$ means that $\hat{H} \cap K \subset K$ and the document does not contain all keywords in K.

Next, the program F computes

$$\mathcal{E}_{pk}(c_0) = \mathcal{E}_{pk}((\gamma_0 \oplus 1) \bigwedge_{w_i \in \hat{H}} (c_{i0} \oplus 1)) \tag{6.2}$$

$$= (\mathcal{E}_{pk}(\gamma_0) + \mathcal{E}_{pk}(1)) \prod_{w_i \in \hat{H}} (\mathcal{E}_{pk}(c_{i0}) + \mathcal{E}_{pk}(1)).$$

If $c_0 = 1$, then $\gamma_0 = 0$ and $c_{i0} = 0$ for all $w_i \in \hat{H}$. As discussed above, $\gamma_0 = 0$ means $\hat{H} \cap K = K$ while $c_{i0} = 0$ for all $w_i \in \hat{H}$ means $f(w_i) \geq t_i$ for all $w_i \in \hat{H}$. It is obvious that $f(w_i) \geq 0$ for all $w_i \notin K$. According to Lemma 6.7, M is a matching document.

If $c_0 = 0$ and $\gamma_0 = 1$, M does not contain all keywords in K. According to Lemma 6.7, M is not a matching document. If $c_0 = 0$ and $\gamma_0 = 0$, M does contain all keywords in K, but there exists i such that $f(w_i) < t_i$. According to Lemma 6.7, M is not a matching document.

The rest of the algorithm and the buffer decryption algorithm are the same as the disjunction threshold query.

The correctness of the conjunctive threshold query can be proved in the same way as we prove the correctness of the disjunctive threshold query.

6.5.3   Complement Threshold Query

There are two complement constructions for private threshold queries based on keyword frequency. They are the disjunctive complement and the conjunctive complement.

6.5.3.1   Disjunctive Complement

Formally, a disjunctive complement threshold query over keywords $K = \{k_1, k_2, \cdots, k_n\}$ can be expressed as

$$\mathcal{Q}_K = (f(k_{i_1}) \geq t_{i_1}) \vee \cdots \vee (f(k_{i_{n_1}}) \geq t_{i_{n_1}})$$

$$\vee \neg (f(k_{j_1}) \geq t_{j_1}) \vee \cdots \vee \neg (f(k_{j_{n_2}}) \geq t_{j_{n_2}})$$

$$= (f(k_{i_1}) \geq t_{i_1}) \vee \cdots \vee (f(k_{i_{n_1}}) \geq t_{i_{n_1}})$$
$$\vee (f(k_{j_1}) < t_{j_1}) \vee \cdots \vee (f(k_{j_{n_2}}) < t_{j_{n_2}}),$$

where $\neg$ stands for complement (i.e., negation), $\{k_{i_1}, \cdots, k_{i_{n_1}}, k_{j_1}, \cdots, k_{j_{n_2}}\} = K$ and $n_1 \geq 0$, $n_2 \geq 0$. It is easy to see

Lemma 6.8 ([23]). *Given a document* M, *a conjunctive complement query* $\mathcal{Q}_K(M) = 1$ *if and only if there exists* l *such that* $f(k_{i_l}) \geq t_{i_l}$ *or* $f(k_{j_l}) < t_{j_l}$.

The construction for the conjunctive complement query is composed of **KeyGen**, **FilterGen**, **FilterExec**, and **BufferDec**, where **KeyGen** and **BufferDec** are the same as the disjunctive threshold query described in Sect. 6.5.1.

Filter Program Generation

FilterGen($D, \mathcal{Q}_K, pk$): This algorithm outputs a filter program F, which consists of the public dictionary $D = \{w_1, w_2, \cdots, w_{|D|}\}$, disjunctive complement sign (denoted as 10), an array of ciphertexts $\hat{D} = \{\hat{w}_1, \hat{w}_2, \cdots, \hat{w}_{|D|}\}$, where $\hat{w}_i = \mathcal{E}_{pk}(t_i)$ and

$$t_i = \begin{cases} \text{frequency threshold for } k_j & \text{if } w_i = k_j \in K \\ 2^d - 1 & \text{if } w_i \notin K \end{cases}$$

and an array of ciphertexts $\hat{D}' = \{\hat{w}'_1, \hat{w}'_2, \cdots, \hat{w}'_{|D|}\}$, where $\hat{w}'_i = \mathcal{E}_{pk}(s_i)$ and

$$s_i = \begin{cases} 1 & \text{if } w_i \in \{k_{j_1}, \cdots, k_{j_{n_2}}\} \\ 0 & \text{otherwise} \end{cases}$$

Remark. The encryptions of $s_1, s_2, \cdots, s_n$ are used to indicate the complement positions in $\mathcal{Q}_K$ in private.

Filter Program Execution

FilterExec(S, F, pk, m): This algorithm outputs an encrypted buffer $\mathbb{B}$ keeping up to m matching documents.

The algorithm is the same as the filter program execution in the disjunctive threshold query described in Sect. 6.5.1. except that F computes

$$\mathcal{E}_{pk}(c_0) = \mathcal{E}_{pk}\left(\bigvee_{w_i \in \hat{H}} (c_{i0} \oplus 1 \oplus s_i) \right) \tag{6.3}$$

on the basis of homomorphic properties described in Sect. 6.3.1.

If $c_0 = 1$, then there exists l such that $c_{l0} \oplus 1 \oplus s_l = 1$ (i.e., $c_{l0} \oplus s_l = 0$). If $w_l \in \{k_{i_1}, \cdots, k_{i_{n_1}}\}$, then $s_l = 0$ and thus $c_{l0} = 0$, which means that $f(w_l) \geq t_l$. If $w_l \in \{k_{j_1}, \cdots, k_{j_{n_2}}\}$, then $s_l = 1$ and thus $c_{l0} = 1$, which means that $f(w_l) < t_l$. If $w_l \notin K$, then $s_l = 0$ and thus $c_{l0} = 0$, which means that $f(w_l) \geq t_l = 2^d - 1$.

It is impossible and this event never occurs when $c_0 = 1$. According to Lemma 6.8, M is a matching document when $c_0 = 1$.

If $c_0 = 0$, then $c_{l0} \oplus 1 \oplus s_l = 0$ (i.e., $c_{l0} \oplus s_l = 1$) for all $w_l \in M \cap D$. If $w_l \in \{k_{i_1}, \cdots, k_{i_{n_1}}\}$, then $s_l = 0$ and thus $c_{l0} = 1$, which means that $f(w_l) < t_l$. If $w_l \in \{k_{j_1}, \cdots, k_{j_{n_2}}\}$, then $s_l = 1$ and thus $c_{l0} = 0$, which means that $f(w_l) \geq t_l$. According to Lemma 6.8, M is not a matching document when $c_0 = 0$.

Remark. A disjunctive complement query becomes a disjunctive query if letting $s_i = 0$ for all i. In addition, if letting $s_i = 1$ for all i, a disjunctive complement query becomes

$$\mathcal{Q}_K = (f(k_1) < t_1) \vee (f(k_2) < t_2) \vee \cdots \vee (f(k_n) < t_n).$$

6.5.3.2   Conjunctive Complement

Formally, a conjunctive complement threshold query over keywords $K = \{k_1, k_2, \cdots, k_n\}$ can be expressed as

$$
\begin{aligned}
\mathcal{Q}_K &= (f(k_{i_1}) \geq t_{i_1}) \wedge \cdots \wedge (f(k_{i_{n_1}}) \geq t_{i_{n_1}}) \\
&\quad \wedge \neg(f(k_{j_1}) \wedge t_{j_1}) \vee \cdots \wedge \neg(f(k_{j_{n_2}}) \geq t_{j_{n_2}}) \\
&= (f(k_{i_1}) \geq t_{i_1}) \wedge \cdots \vee (f(k_{i_{n_1}}) \geq t_{i_{n_1}}) \\
&\quad \wedge (f(k_{j_1}) < t_{j_1}) \wedge \cdots \vee (f(k_{j_{n_2}}) < t_{j_{n_2}}),
\end{aligned}
$$

where $\neg$ stands for complement (i.e., negation), $\{k_{i_1}, \cdots, k_{i_{n_1}}, k_{j_1}, \cdots, k_{j_{n_2}}\} = K$ and $n_1 \geq 0, n_2 \geq 0$. It is easy to see

Lemma 6.9 ([23]). *Given a document M, a conjunctive complement query $\mathcal{Q}_K(M) = 1$ if and only if, for any $k_l \in \{k_{i_1}, k_{i_2}, \cdots, k_{i_{n_1}}\}$, $f(k_l) \geq t_l$, and for any $k_l \in \{k_{j_1}, k_{j_2}, \cdots, k_{j_{n_2}}\}$, $f(k_l) < t_l$.*

The construction for the conjunctive complement query is composed of KeyGen, FilterGen, FilterExec, and BufferDec, where KeyGen and BufferDec are the same as the disjunctive threshold query described in Sect. 6.5.1.

Filter Program Generation
FilterGen($D, \mathcal{Q}_K, pk$): This algorithm outputs a filter program F, which consists of the public dictionary $D = \{w_1, w_2, \cdots, w_{|D|}\}$, conjunctive complement sign (denoted as 11), an array of ciphertexts $\hat{D} = \{\hat{w}_1, \hat{w}_2, \cdots, \hat{w}_{|D|}\}$, where $\hat{w}_i = \mathcal{E}_{pk}(t_i)$ and

$$
t_i = \begin{cases} \textit{frequency threshold for } k_j & \text{if } w_i = k_j \in K \\ 0 & \text{if } w_i \notin K \end{cases}
$$

and an array of ciphertexts $\hat{D}' = \{\hat{w}'_1, \hat{w}'_2, \cdots, \hat{w}'_{|D|}\}$, where $\hat{w}'_i = \mathcal{E}_{pk}(s_i)$ and

$$s_i = \begin{cases} 1 & \text{if } w_i \in \{k_{j_1}, \cdots, k_{j_{n_2}}\} \\ 0 & \text{otherwise} \end{cases}$$

Filter Program Execution

FilterExec(S, F, pk, m): This algorithm outputs an encrypted buffer $\mathbb{B}$ keeping up to m matching documents.

The algorithm is the same as the filter program execution in the conjunctive threshold query described in Sect. 6.5.2. except that F computes

$$\mathcal{E}_{pk}(c_0) = \mathcal{E}_{pk}((\gamma_0 \oplus 1) \bigwedge_{w_i \in \hat{H}} (c_{i0} \oplus 1 \oplus s_i)) \tag{6.4}$$

$$= (\mathcal{E}_{pk}(\gamma_0) + \mathcal{E}_{pk}(1)) \prod_{w_i \in \hat{H}} (\mathcal{E}_{pk}(c_{i0}) + \mathcal{E}_{pk}(1) + \hat{w}'_i).$$

according to homomorphic properties described in Sect. 6.3.1.

If $c_0 = 1$, then $\gamma_0 = 0$ and $c_{l0} \oplus 1 \oplus s_l = 1$ (i.e., $c_{l0} + s_l = 0$) for all $w_l \in \hat{H}$. $\gamma_0 = 0$ means $\hat{H} \cap K = K$. If $w_l \in \{k_{i_1}, \cdots, k_{i_{n_1}}\}$, then $s_l = 0$ and thus $c_{l0} = 0$, which means that $f(w_l) \geq t_l$. If $w_l \in \{k_{j_1}, \cdots, k_{j_{n_2}}\}$, then $s_l = 1$ and thus $c_{l0} = 1$, which means that $f(w_l) < t_l$. According to Lemma 6.9, M is a matching document when $c_0 = 1$.

If $c_0 = 0$ and $\gamma_0 = 1$, M does not contain all keywords in K. According to Lemma 6.9, M is not a matching document. If $c_0 = 0$ and $\gamma_0 = 0$, M does contain all keywords in K, but there exists l such that $c_{l0} \oplus 1 \oplus s_l = 0$ (i.e., $c_{l0} \oplus s_l = 1$). If $w_l \in \{k_{i_1}, \cdots, k_{i_{n_1}}\}$, then $s_l = 0$ and thus $c_{l0} = 1$, which means that $f(w_l) < t_l$. If $w_l \in \{k_{j_1}, \cdots, k_{j_{n_2}}\}$, then $s_l = 1$ and thus $c_{l0} = 0$, which means that $f(w_l) \geq t_l$. According to Lemma 6.9, M is a matching document when $c_0 = 0$.

Remark. A conjunctive complement query becomes a conjunctive query if letting $s_i = 0$ for all i. In addition, if letting $s_i = 1$ for all i, a disjunctive complement query becomes

$$\mathcal{Q}_K = (f(k_1) < t_1) \wedge (f(k_2) < t_2) \wedge \cdots \wedge (f(k_n) < t_n).$$

6.5.4 Generic Threshold Query

By combining the above basic constructions for private threshold queries based on keyword frequency, we present the construction for a generic threshold query without asymptotically increasing the program size as follows.

Assume that D is the public dictionary of potential keywords and $\mathcal{Q}_{K_i}^{(i)}$ ($i = 1, 2, \cdots, \lambda$) stands for a disjunctive, or conjunctive, or complement query over keywords $K_i \subset D$; we consider a generic threshold query

$$\Phi(\mathcal{Q}_{K_1}^{(1)}, \mathcal{Q}_{K_2}^{(2)}, \cdots, \mathcal{Q}_{K_\lambda}^{(\lambda)}),$$

where operators in Φ belong to $\{\vee, \wedge, \oplus\}$ and $K_i \cap K_j$ for any i and j is not necessary to be empty.

The construction for the generic threshold query over keywords K_i ($i = 1, 2, \cdots, \lambda$) is composed of KeyGen, FilterGen, FilterExec, and BufferDec, where KeyGen and BufferDec are the same as the threshold queries described in Sect. 6.5.1.

Filter Program Generation

FilterGen($D, \mathcal{Q}_{K_1}^{(1)}, \mathcal{Q}_{K_2}^{(2)}, \cdots, \mathcal{Q}_{K_\lambda}^{\lambda}, pk$): This algorithm outputs a filter program F, which consists of $\{F_1, F_2, \cdots, F_\lambda\}$ where $F_i = $ FilterGen($D, \mathcal{Q}_{K_i}^{(i)}, pk$).

Filter Program Execution

FilterExec(S, F, pk, m): This algorithm outputs an encrypted buffer $\mathbb{B}$ keeping up to m matching documents. Upon receiving an input document $M = (m_1 m_2 \cdots m_\ell)_b$ from the stream S, the program F proceeds with the following steps:

1. The program F runs the programs F_i to compute $\mathcal{E}_{pk}(c_0^{(i)})$ based on Eq. (6.1)–(6.4).
2. The program F computes

$$\mathcal{E}_{pk}(c_0) = \mathcal{E}_{pk}(\Phi(c_0^{(1)}, c_0^{(2)}, \cdots, c_0^{(\lambda)})).$$

according to homomorphic properties described in Sect. 6.3.1.

If $c_0 = 1$, M is a matching document. If $c_0 = 0$, M is not a matching document.

The rest of the construction is the same as FilterExec of the disjunction threshold query described in Sect. 6.5.1.

Remark. All kinds of private threshold queries based on keyword frequency can be expressed as $\Phi(\mathcal{Q}_{K_1}^{(1)}, \mathcal{Q}_{K_2}^{(2)}, \cdots, \mathcal{Q}_{K_\lambda}^{(\lambda)})$, where $\mathcal{Q}_{K_i}^{(i)}$ is either disjunctive, conjunctive, or complement threshold query, and operators in Φ belong to $\{\vee, \wedge, \oplus\}$. Therefore, the solution supports arbitrary private threshold queries.

6.6   Performance Analysis

In the disjunctive construction (Sect. 6.5.1), the client can pre-generates the public/private key pair. In addition, the client needs to encrypt the frequency of each classified keyword in the phase of the filter program generation and to decrypt the buffer $\mathbb{B}$ to retrieve the matching documents after the buffer returns. If one does not consider the key generation, the total computation complexity of the client is $O(d|D|)$ encryptions to generate the program F and $O(m\ell)$ decryptions to retrieve the matching documents from the buffer, where $|D|$ is the number of words in the dictionary D, 2^d is the maximal number of words contained in each document, ℓ is the number of bits of each document, and m is the maximal number of matching documents in the buffer.

After receiving the filter program F, the server processes each document M_i in three steps. We assume $\mu = |M_i \cap D|$. At first, the server needs to compute $\mathcal{E}_{pk}(c_0)$. The computation complexity of the first step is $O(\mu d)$ encryptions to encrypt μ frequencies with d bits, $O(\mu)$ homomorphic additions of integers with d bits, $O(\mu)$ homomorphic multiplications of bits, and $O(\mu)$ homomorphic additions of bits (please refer to Eq. (6.1)). Then, the server needs to add M_i into the buffer $\mathbb{B}$ if M_i is a matching document or add 0 into the buffer otherwise. The computation complexity of the second step is $O(m\ell^2)$ homomorphic multiplications of bits and $O(m\ell^2)$ homomorphic addition of bits. At last, the server needs to update the buffer base $\mathbb{G}$. The computation complexity of the third step is $O(m)$ homomorphic multiplications of bits and $O(1)$ homomorphic addition of integers with m bits.

The communication complexity of the disjunctive construction is $O(d|D||C|)$ bits for the query and $O(m\ell|C|)$ bits for response, where $|C|$ is the size of the ciphertext.

Unlike the disjunctive construction, the conjunctive construction (Sect. 6.5.2) needs to compute $\mathcal{E}_{pk}(\gamma_0)$ and then $\mathcal{E}_{pk}(c_0)$. The computation complexity for the server to compute $\mathcal{E}_{pk}(\gamma_0)$ is $O(|D|)$ homomorphic additions of integers. Although the two constructions computes $\mathcal{E}_{pk}(c_0)$ with two different equations (please refer to Eqs. (6.1) and (6.2)), their complexities for this computation are almost the same.

The disjunctive complement construction (Sect. 6.5.3.1) is different from the disjunctive construction in two ways. The query contains an extra array of ciphertexts to indicate the complement positions in private and the server computes $\mathcal{E}_{pk}(c_0)$ with Eq. (6.3), which is different from Eq. (6.1). The differences do not affect the computation complexity of the server, but the computation complexity of the client is increased by $O(|D|)$ encryptions of bits and the communication complexity is increased by $O(|D||C|)$ bits on the basis of the performance of the disjunctive construction.

Similarly, the conjunctive complement construction (Sect. 6.5.3.2) is different from the conjunctive construction in two ways. The differences do not change the computation complexity of the server, but the computation complexity of the client is increased by $O(|D|)$ encryptions of bits and the communication complexity is increased by $O(|D||C|)$ bits on the basis of the conjunctive complement construction.

The performance of the generic construction (Sect. 6.5.4) depends on the performance of the underlying basic constructions.

The performance comparison of the threshold query protocols can be summarized in Table 6.2, where enc. and dec. stand for encryption and decryption of bit, add. and multi. denote the homomorphic addition and multiplication of bits, and ADD. represents the homomorphic addition of integers.

Table 6.2 Performance comparison

Protocols	Comp. complexity (client)	Comp. complexity (sever, M_i)	Comm. complexity
Disjunctive	$O(d|D|)$ enc. +$O(m\ell)$ dec.	$O(\mu d)$ enc. +$O(\mu)$ ADD. +$O(m\ell^2 + \mu)$ multi. +$O(m\ell^2 + \mu)$ add.	$O(d|D||C|)$ +$O(m\ell)$
Conjunctive	Same as disjunctive	Disjunctive +$O(|D|)$ ADD.	Same as disjunctive
Disjunctive complement	Disjunctive +$O(|D|)$ enc.	Same as disjunctive	Disjunctive +$O(|D||C|)$
Conjunctive complement	Disjunctive +$O(|D|)$ enc.	Same as conjunctive	Disjunctive +$O(|D||C|)$

6.7 Conclusion and Discussion

On the basis of the state of the art fully homomorphic encryption techniques, we describe constructions for disjunctive, conjunctive, and complement threshold queries based on keyword frequency and then the construction for the generic threshold query based on keyword frequency given by Yi et al. [23]. These protocols are semantically secure as long as the underlying fully homomorphic encryption scheme is semantically secure.

The construction for disjunctive threshold query is able to search for documents containing at least one of a set of keywords as [1, 2, 15, 16] by letting the threshold $t_i = 1$ for keyword $k_i \in K$. The construction for generic threshold query can search for documents M such that $(M \cap K_1 \neq \emptyset) \wedge (M \cap K_1 \neq \emptyset)$ as [15, 16] by letting $Q_{K_1}^{(1)}$ and $Q_{K_2}^{(2)}$ be two disjunctive threshold queries with the threshold $t_i = 1$ for keyword $k_i \in K$ and $\Phi(Q_{K_1}^{(1)}, Q_{K_2}^{(2)}) = Q_{K_1}^{(1)} \wedge Q_{K_2}^{(2)}$. Therefore, their solutions are special cases of the protocols given by Yi et al. [23]

To improve the performance of the constructions, the ciphertext of a bit in the final stage of filter program execution can be compressed or post-processed as [6]. In this case, the ciphertext of a bit can have the same size as an RSA modulus asymptotically.

Theoretically, any search criteria can be constructed with fully homomorphic encryption scheme in private searching on streaming data. Even if so, different queries will need different constructions. As long as the underlying fully homomorphic encryption scheme is practical, the protocols will be practical. So far, fully homomorphic encryption schemes are impractical for many applications according to [11], because ciphertext size and computation time increase sharply as one increases the security level. Recently, many research efforts have been devoted to construct efficient fully homomorphic encryption schemes, such as the ones by

[4, 5, 21]. We believe that the protocols for private threshold queries based on keyword frequency will be made practical with the performance improvement of fully homomorphic encryption techniques in the future.

Privacy is gaining increasingly higher attention, and future computing paradigms, e.g., cloud computing, will only become viable if privacy of users is thoroughly protected. For example, Google Alerts is a service offered by Google which notifies its users by e-mail, or as a feed, about the latest Web and news pages of their choice. As in the case of the AOL search data leak, it is not hard to imagine queries which could be privacy sensitive. With the private searching solutions, it is possible for a user to make a filtering program according to the frequencies of some classified keywords and submit it to Google, which executes the program on all latest Web and news pages. The program can notify to the user its discovery according to the search criteria specified by the user. While the program is executed by Google, the search criteria of the user can be kept confidential to Google.

References

1. J. Bethencourt, D. Song, B. Water, New construction and practical applications for private streaming searching, in *Proceedings of SP'06*, 2006, pp. 132–139
2. J. Bethencourt, D. Song, B. Water, New techniques for private stream searching. ACM Trans. Inform. Syst. Secur. **12**(3), 1–32 (2009)
3. D. Boneh, E. Goh, K. Nissim, Evaluating 2-DNF formulas on ciphertexts, in *Proceedings TCC'05*, 2005, pp. 325–341
4. Z. Brakerski, C. Gentry, V. Vaikuntanathan. (Leveled) fully homomorphic encryption without bootstrapping, in *Proceedings of ITCS'12*, 2012, pp. 309–325
5. Z. Brakerski, V. Vaikuntanathan. Efficient fully homomorphic encryption from (standard) LWE, in *Proceedings of FOCS'11*, 2011, pp. 97–106
6. M. van Dijk, C. Gentry, S. Halevi, V. Vaikuntanathan. Fully homomorphic encryption over the integers, *Proceedings of EUROCRYPT'10* 2010 pp. 24–43
7. C. Gentry. Fully Homomorphic Encryption Scheme. PhD thesis, Stanford University, 2009
8. C. Gentry, Fully homomorphic encryption using ideal lattices, in *Proceedings of STOC'09*, 2009, pp. 169–178
9. C. Gentry, Computing arbitrary functions of encrypted data. Commun. ACM **53**(3), 97–105 (2010)
10. C. Gentry, Toward basing fully homomorphic encryption on worst-case hardness, in *Proceedings of CRYPTO'10*, 2010, pp. 116–137
11. C. Gentry, S. Halevi, Implementing Gentry's fully-homomorphic encryption scheme, in Proceedings nof EUROCRYPT'11, 2011, pp. 129–148
12. D. Harris, S. Harris, *Digital Design and Computer Architecture* (Morgan Kaufmann Publishers, Massachusetts, 2007)
13. D.J. Lilja, S.S. Sapatnekar, *Designing Digital Computer Systems with Verilog* (Cambridge University Press, Cambridge, 2005)
14. S. Ling, C.P. Xing, *Coding Theory: A First Course* (Cambridge Press, Cambridge, 2004)
15. R. Ostrovsky, W. Skeith, Private searching on streaming data, in *Proceedings of CRYPTO'05*, 2005, pp. 223–240
16. R. Ostrovsky, W. Skeith, Private searching on streaming data. J. Cryptol. **20**(4), 397–430 (2007)
17. R. Ostrovsky, W. Skeith, Algebraic lower bounds for computing on encrypted data. Electronic Colloquium on Computational Complexity (ECCC), Report No. 22, 2007

18. P. Paillier, Public key cryptosystems based on composite degree residue classes, in *Proceedings of EUROCRYPT'99*, 1999, pp. 223–238
19. B. Parhami, *Computer Arithmetic: Algorithms and Hardware Designs*, 2nd edn. (Oxford University Press, Oxford, 2010)
20. N. Smart, F. Vercauteren, Fully homomorphic encryption with relatively small key and ciphertext sizes, in *Proceedngs of PKC'10*, 2010, pp. 420–443
21. D. Stehle, R. Steinfeld, Faster fully homomorphic encryption, in *Proceedings of ASIACRYPT'10*, 2010, pp. 377–394
22. J.F. Wakerly, *Digital Design Principles & Practices*, 3rd edn. (Prentice Hall, New Jersey, 2000)
23. X. Yi, E. Bertino, J. Vaidya, C. Xing, Private searching on streaming data based on keyword frequency. IEEE Trans. Dependable Sec. Comput. **11**(2), 155–167 (2014)
24. X. Yi, C.P. Xing, Private (t, n) threshold searching on streaming data, in *Proceedings of PASSAT'12*, 2012, pp. 676–683

CPSIA information can be obtained at www.ICGtesting.com
Printed in the USA
LVOW05s1542100715

445787LV00004B/6/P

9 783319 122281